Wetlands of the Adirondacks: Herbaceous Plants and Aquatic Plants

Meiyin Wu & Dennis Kalma

Trafford
PUBLISHING®

Order this book online at www.trafford.com
or email orders@trafford.com

Most Trafford titles are also available at major online book retailers.

Printed in the United States of America.

ISBN: 978-1-4269-5843-4 (sc)
ISBN: 978-1-4269-6062-8 (hc)
ISBN: 978-1-4269-5844-1 (e)

Library of Congress Control Number: 2011902742

Trafford rev. 02/17/2011

 www.trafford.com

North America & International
toll-free: 1 888 232 4444 (USA & Canada)
phone: 250 383 6864 ♦ fax: 812 355 4082

Table of Contents

Acknowledgements

We want to thank agencies that provided funding and/or support: the United States Environmental Protection Agency, the United States Department of Agriculture Natural Resources Conservation Service the New York State Department of Environmental Conservation, the New York State Adirondack Park Agency, the Nature Conservancy, Ausable River Association, Boquet River Association, State University of New York College at Plattsburgh, and Montclair State University. Much appreciation goes to the many landowners who have allowed us access to the wetlands on their lands. Special thanks to Dr. Kenneth Adams and Mr. Dan Spada who shared their knowledge and expertise. We thank Mr. Gabriel Jimenez for his technical assistance. A final note of deepest love and gratitude goes to our families—Alice, Allen, Eddie, Michael, and Meiyin's parents, whom without their love, patience and support, this book could never be finished.

Introduction

This book is the first of two books on identification of wetland plants in the Adirondack region. Due to page limitation, we cannot possibly include every wetland plant in the Adirondack region; we tried to include the most common ones. Although not true wetland plants, some species are commonly found within Adirondack wetlands; therefore, those species were selected to be included intentionally. The two books begin by splitting the species into five major groups:

Book 1:
    Ferns and Allies
    Woody Plants
    Grass-like Plants

Book 2:
    Herbaceous Plants
    Aquatic Plants

While some species might be easy to identify, others may belong to a group of species where identification can be a difficult technical task and requires specific botanical information. Every effort has been made to minimize the amount of terminology used in these books. Where possible, the rather bewildering descriptive terms used in technical manuals have been replaced by plain English translations. We hope these two books will be useful to anyone without botanical training. Some familiarity with the names of parts of plants is, however, necessary. The following section gives a minimal introduction to the structure of plants and the terminology involved to describe them. Some specialized structures, found only in certain groups of plants, are described in the introduction to those groups.

Identification by Field Guide versus Keys:

Some species are so distinctive that it is a waste of time to construct and use a key to distinguish among them. For example, a jack-in-the-pulpit is almost impossible to confuse with anything else in Adirondack wetlands, so it makes sense to just look at the individual accounts of the species in the subgroup the general key leads you to and choose among them. Other species are more difficult (for example, the sedges) and have such a mix characteristics that they are difficult to distinguish among without a key. In these cases a key will be helpful and will lead you directly to the identity of a species. Or in some cases, the key will lead you to a list of species; you can then look at the individual accounts of these species and determine the appropriate one from among them.

Keys:

A key is like the game "Twenty Questions". At each level you are asked to choose between two answers to a question or between two statements. Depending on your choice, you are led to other questions. Each question in turn eliminates some possible species. After a series of questions a final question leads you to the name of the species (or a group of species).

In these books there are always two, and only two, possible choices at each level of the key. The alternate choices are called a "couplet". At the left hand side of the page is a number followed by "a" or "b" that identifies the couplet, e.g. 1a) & 1b). Each half of the couplet is followed by statements between which you must choose. At the end of the line of each half of the couplet is a number that points you to the next set of couplets to choose between. If the choice in the couplet is the end point of that branch of the key, then the line ends in the name of the species or in the name of a group to which you proceed for another key. As an example, for the identification of the major groups, as described above, a very simple key would be constructed as follows:

1a) Plants simple: fern-like or horsetail or quillworts... Ferns & Allies (*Go to Book 1 page 7*)
1b) Others ... 2

2a) Plants with woody stems ... Woody Plants (*Go to Book 1 page 23*)
2b) Plants without woody stems ... 3

3a) Plants with grass-like stems and leaves ... Grass-like Plants (*Go to Book 1 page 110*)
3b) Plants not grass-like ... 4

4a) Plants with leaves entirely underwater, floating on the surface, or plants free-floating ... Aquatic Plants (*Go to Book 2 page 149*)
4b) Plants without leaves entirely underwater, floating on the surface, or plants free-floating ... Herbaceous Plants (*Go to Book 2 page 7*)

Flower Anatomy:

The structure of the flower is probably the most important characteristic biologists use to determine the evolutionary relationships among the plants. Because of this most keys for identification rely heavily on the flowers. Although they may not always be available for use, field biologists usually consider them the "gold standard" for identification.

The structure of a typical flower is shown in the illustration:

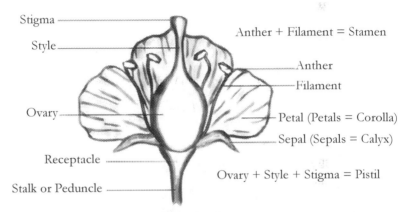

The ovary, stigma, and style, the female reproductive parts, are collectively known as the pistil. The filament and anther, the male reproductive parts, are collectively known as the stamen. The accessory structures, the petals (collectively known as the corolla) and the sepals (collectively known as the calyx) may or may not be present; the petals and sepals together are known as the perianth or, in some groups, as the tepals. The flower may have radial symmetry (the symmetry of a circle – looking down from the top, the sides are the same no matter where you divide it) or may have bilateral symmetry (there are definite right and left sides, same as with our own bodies). Different keys use different words for these terms; for example radial symmetry is often known as "rotate" or "symmetrical", while bilateral symmetry is often known as "zygomorphic".

The petals or the sepals may fuse all along their length, or along part of their length, to give a tubular corolla or tubular calyx. In the diagram above the ovary sits on top of the receptacle – a superior ovary. The receptacle may grow up and around the ovary so it appears as if the calyx and corolla are attached above the ovary – an inferior ovary. In many flowers, such as the one pictured here, the female (pistil) and male (stamen) parts are found in the same flower. These flowers are considered perfect. In other cases, imperfect or unisexual flowers, the male parts are found in some flowers, staminate flowers, and the female parts are found in other flowers, pistillate flowers. Staminate and pistillate flowers may be found on the same plant (dioecious plants) or they may be found on different plants (monoecious plants).

Flowers may be single or they may be grouped in inflorescences. The various arrangements have different names:

1. **Spike**: In a spike the the individual flowers are attached directly to a central stem (the rachis).
2. **Raceme**: A raceme is similar to a spike but the individual flowers are at the ends of side branches of the rachis.
3. **Panicle**: A panicle is similar to a raceme but the side branches are themselves branched.
4. **Corymb**: A corymb is like a raceme but the length of the branches varies so that all of the flowers are in a more or less flat-topped inflorescence. In a compound corymb each of the side branches is branched.
5. **Umbel**: In an umbel all of the flower stalks arise at a single point at the top of the stem. The inflorescence may be flat-topped or rounded. In a compound umbel each of the branches is branched.
6. **Head**: In a head all of the flowers arise from a small "head".

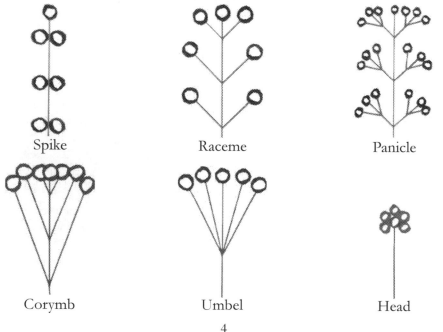

Spike   Raceme   Panicle

Corymb   Umbel   Head

Vegetative Structures:

The leaves and their arrangement are also used in identification. Some perennial plants have underground rhizome systems which extend horizontally and send out new roots and shoots year after year. Parasitic plants may have "No Leaves". In a few other plants the leaves may disappear either before or after the flowers are present; these are referred to as "No Leaves At Flowering". When leaves are present, they may arise directly from the ground, "Leaves Basal", or they may be on the stem. In some herbaceous plants the distinction of whether the leaves are truly basal or whether they just arise near the base of the stem is problematic; in these plants we have placed the species in both groupings – whichever choice you make should lead you to the correct answer. If they are on the stem they may be arranged in pairs or whorls opposite one another, "Leaves Opposite or Whorled", or they may be arranged one after the other on alternate sides of the stem, "Leaves Alternate". The leaves may consist of a single undivided blade, "Leaves Simple", or the blade may be subdivided into leaflets, "Leaves Compound". The leaf (or leaflet) may have a smooth edge ("Entire") or it may have teeth ("Toothed") or it may have lobes ("Lobed"). If an indentation in the blade of the leaf extends only part way to the central vein, the leaf considered is lobed; if the indentation extends to the central vein, so that there is a stretch of vein without blade tissue along it, the leaf is subdivided into leaflets and is considered compound.

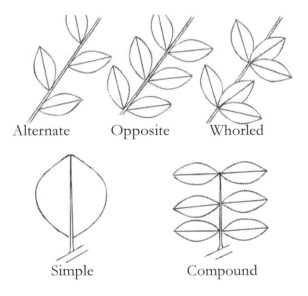

Alternate   Opposite   Whorled

Simple   Compound

5

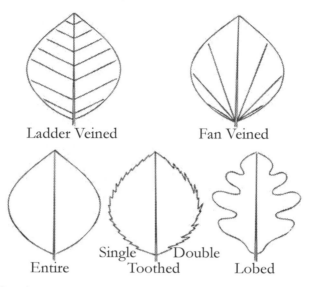

Ladder Veined   Fan Veined

Entire   Single Double   Lobed
            Toothed

Collecting Specimens:

       If a species cannot be identified in the field, specimens might
be collected to verify identifications. Particularly in the case of grasses-
like species and certain other groups, where identification of species is
often difficult and depends on microscopic characters. When collecting
specimens of a tree, collect a short segment of branch with attached leaves
from and document its height, the texture of its bark, and so on. When
collecting specimens of the grass-like species, care should be taken to
collect a portion of the roots as well as the stems and seed heads as these
are often critical characters. Specimens collected should be placed in plastic
bags and kept in a cool location (refrigerate if necessary to hold overnight).
If the specimen seems to be rare it should not be collected. Many biologists
use the "rule of 20" as a guideline for collecting unidentified specimens:
if there are more than 20 plants in the area it is appropriate to collect one
for identification. Orchids, however, should never be collected, even if they
seem abundant.
If the plants are considered too rare to collect, the best solution is to take
several pictures of the plant. Get up as close as possible to the plant and
take photographs from several angles and of different parts of the plant.

Herbaceous Plants

If flowers are present, they are the fastest, most convenient way to identify herbaceous plants. However, flowers on any species are likely to be present for only a short portion of the field season. Additionally, in some species the flowers are so small that it is difficult to identify without a dissecting microscope. Therefore, it is necessary to identify herbaceous plants by its foliage characteristics. The following table provide possible identifications based on a few foliage characteristics; positive identification depends on matching other characters described in the individual species sheets which follow (listed alphabetically by scientific name).

First, decide if the plant has basal leaves only. If it does not, decide whether the stem leaves are arranged alternately, or are arranged oppositely and/or in whorls. Second, decide if the plant has compound leaves. This is a bit trickier than in woody plants where just what is a leaf is more obvious. If an indentation in the blade of the leaf extends only part way to the central vein, the leaf considered is lobed; if the indentation extends to the central vein, so that there is a stretch of vein without blade tissue along it, the leaf is subdivided into leaflets and is considered compound. In some species it will be difficult to determine if the leaves are compound or simple - you will have to try both options. If the leaf is not compound decide if the margin is toothed, lobed, or entire. You can look through the possible species in the pages following these charts to determine the identity of the specimen. If nothing really seems to work, you may have found a new species.

Basal Leaves Only

Compound leaves	Toothed/lobed leaves	Entire leaves
Arisaema dracontium	*Mitella nuda*	*Acorus americanus*
Arisaema triphyllum	*Packera schweinitziana*	*Alisma subcordatum*
Coptis trifolia	*Pyrola asarifolia*	*Asarum canadense*
Elatine minima	*Sanguinaria canadensis*	*Drosera rotundifolia*
Eriocaulon aquaticum	*Tiarella cordifolia*	*Hieracium caespitosum*
Waldensteia fragarioides	*Viola affinis*	*Iris pseudacorus*
	Viola blanda	*Iris versicolor*
		Liparis loeseli
		Littorella uniflora
		Lobelia dortmanna
		Peltandra virginica
		Pontederia cordata
		Ranunculus abortive
		Sagittaria latifolia
		Sarracenia purpurea
		Sisyrinchium angustifolium
		Sparganium androcladum
		Sparganium erectum
		Sparganium eurycarpum
		Spiranthes romanzoffiana
		Symplocarpus foetidus
		Vesrbacum Thapsus
		Xyris montana

Alternate Leaves

Compound leaves

Agrimonia gryosepala
Amphicarpa bracteata
Apios americana
Cardamine pensylvanica
Cicuta bulbifera
Clematis virginiana1
Comarum palustre
Conioselinum chinense
Empetrum nigrum
Filipendula rubra
Lathyrus palustris
Menyanthes trifoliate
Ranunculus flabellarus
Ranunculus pensylvanicus
Sium suave
Thalictrum pubescens

Toothed/lobed leaves

Alliaria petiolata
Caltha palustris
Cirsium muticum
Epilobium coloratum
Epilobium strictum
Geum macrophyllum
Hydrocotyl americana
Impatiens capensis
Laportea canadensis
Lobelia cardinalis
Ranunculus acris
Rorippia palustris
Solanum dulcamara
Solidago gigantea
Solidago patula
Solidago scabra
Solidago uliginosa
Symphyotrichum novii-belgi
Symphyotrichum racemosum
Utricularia macrorhiza

Entire leaves

Arethusa bulbosa
Calla palustri
Calopogon tuberosus
Calypso bulbosa
Campanula aparinoides
Cypripedium reginae
Euthamia graminifolia
Gaultheria hispidula
Iris pseudacorus
Iris versicolor
Maianthemum canadensis
Maianthemum stellata
Myosotis scorpioides
Oclemena nemoralis
Platanthera blephariglottis
Pogonia ophioglossoides
Polygonum amphibium
Polygonum cilinode
Polygonum persicaria
Pontederia cordataRumex
verticellatus
Sparganium androcladum
Sparganium erectum
Sparganium eurycarpum
Typha latifolia
Vaccinium macrocarpon
Veratrum viride
Verbascum thapsus

Opposite Leaves

Compound leaves

Anemone quinquefolia
Angelica atropurpurea
Bidens frondosa
Clematis virginiana
Linnaea borealis

Toothed/lobed leaves

Ageratina altissima
Boehemeria cylindrica
Chelone glabra
Chrysosplenium
americanum
Circaea lutetiana
Eupatorium perfoliatum
Gratiola neglecta
Impatiens capensis
Lycopus americana
Lycopus uniflorus
Mentha arvensis
Pilea pumila
Scutellarifa galericulata
Stachys tenuifolia
Urtica dioicia
Verbena hastata

Entire leaves

Agalinis paupercula
Arisaema triphyllum
Asclepias incarnata
Bartonia virginica
Chrysosplenium
americanum
Decodon verticillatus
Doellingeria umbellate
Gentiana andrewsii
Gentianopsis crinite
Hypericum elipticum
Hypericum mutilum
Hypericum perfoliatum
Justicia Americana
Listera cordata
Ludwigia palustris
Lysimachia ciliata
Lysimachia nummularia
Lysimachia terrestris
Lysimachia thrysiflora
Lythrum salicaria
Pycnanthemum tenuifolium
Triadenum virginianum
Veronica scutellaria

Whorled Leaves

<u>Toothed/lobed leaves</u>

Anemone canadensis
Eupatorium maculatum

<u>Entire leaves</u>

Cornus canadensis
Decodon verticillatus
Galium asperellum
Galium palustre
Gentiana andrewsii
Hippuris vulgaris
Lilium canadense
Trillium grandiflorum

Sweetflag
Acorus americanus (Raf.) Raf

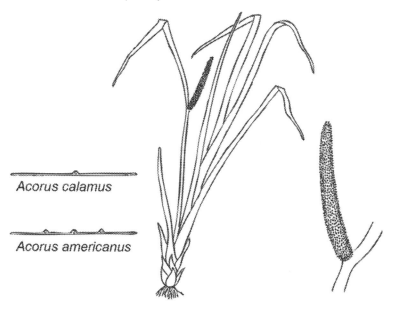

Acorus calamus

Acorus americanus

Plants: up to 150 cm; erect; perennial; sweet spicy smell when crushed
Flowers: 6-parted; yellowish to brown; tiny in size; inflorescence a laterally protruding 5-10 cm long cylindrical spadix
Fruits: brown
Leaves: with 2-6 major raised veins; linear; long and narrow; sheathing base
Habitat: marshes, wet meadows, shallow waters
Similar Species: Calamus, *Acorus calamus*, is an exotic species from Europe and is now more common than the native form; it has a single raised vein in the middle of the leaf

Red Baneberry
Actaea rubra (Aiton) Willd.

Status: Unknown
Ranunculaceae

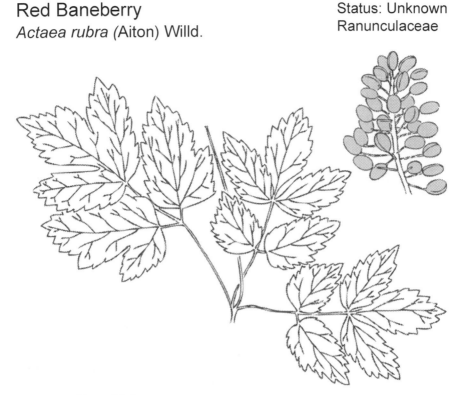

Plant: 30-90 cm high
Flower: single raceme 3-5 cm long on thick stalk; flowers usually in a tight cluster that is somewhat longer than wide; individual flowers with 4-10 small narrow white petals (inconspicuous due to many long white stamens)
Fruit: berries red (occasionally white); stalks thin and long
Leaves: whorl of 3 leaves, each subdivided into leaflets with sharp and irregular teeth
Habitat: woods
Similar Species: most easily distinguished from White Baneberry, A. pachypoda, which has white fruit, flower & fruit clusters that are distinctly longer than wide and longer & thinner flower stalks

13

Smallflower False Foxglove
Agalinis paupercula (A. Gray)

Status: FACW+
Scrophulariaceae

Plants: 15-50 cm tall; erect; annual; dark green, 4 -angled smooth stem
Flowers: 5 parted, pink to purplish pink; 1.2-2.2 cm long, hairy inside; 4 stamens of various length
Fruits: roundish capsule; up to 1.2 cm long
Leaves: opposite, narrow, dark green
Habitat: shoreline; wet meadows, bogs
Similar Species: Slender False Foxglove, *A. tennuifolia*, has smaller flowers (usually 1.0 cm or less) and is usually found in dry woods or fields but may also be in wet areas
aka: *Gerardia paupercula*

White Snakeroot
Ageratina altissima (L.) King & H.E. Robins.

Status: FACU-
Asteraceae

Plant: 0.3-1.5 m
Flower: tiny white flowers on heads arranged in multiple clusters
Leaves: opposite, on definte stalks more than 2 cm long; toothed
Habitat: rich woods, thickets
Similar Species: definite stalks on leaves distinguish this species from closely related *Eupatorium* species
aka: *Eupatorium rugosum*

Water Plantain
Alisma subcordatum Raf.

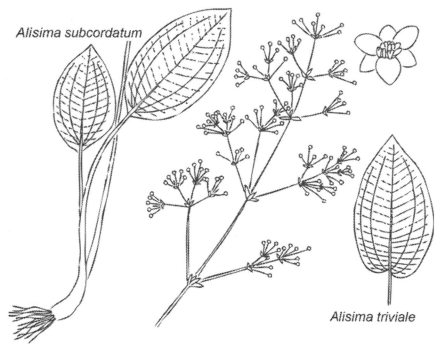

Alisima subcordatum

Alisima triviale

Plants: 10-100 cm tall; erect or floating, perennial
Flowers: 3 parted, white or occasionally pinkish, 7 mm or more across
Leaves: basal, long stalked, elliptical with tapered base; parallel veins
Habitat: shallow water or mud
Similar Species: Northern Water Plantain, *A. triviale*, is very similar but can be distinguished by its smaller (~3 mm) flowers (if present), and the somewhat rounded or heart-shaped base to the leaves

Garlic Mustard
Alliaria petiolata (Bieb.) Cavara & Grande

Status: FACU-
Brassicaceae

Plant: 30-100 cm high
Flower: 4 parted, white, 7-10 mm wide
Fruit: elongate pods erect
Leaves: alternate, triangular to heart-shaped, stalked, toothed; smell of garlic when crushed
Habitat: roadsides, open woods
Similar Species: leaf shape and smell distinctive

Hog Peanut
Amphicarpaea bracteata (L.) Fern.

Plant: delicate twining vine, climbs on other plants
Flower: bilateral, pale purple, violet, or white; narrow, 12-18 mm long, in racemes in the axils of leaves; also - flowers without petals at base of plant
Fruit: aerial flowers produce curved pod with 3-4 beans; ground flower produces a fleshy pod with 1 bean
Leaves: alternate, compound; 3 leaflets, broad, pointed on ends, entire, light green, delicate
Habitat: moist woods
Similar Species: Groundnut, *Apios americana* is a more robust vine with 5 leaflets

Canadian Anemone
Anemone canadensis L.

herbaceous

Status: FACW
Ranunculaceae

Plants: 30-60 cm; perennial; from rhizome
Flowers: mostly 5 parted; white sepals (no petals) 2.5-3.8 cm across around a gold center; solitary, on long stalk
Leaves: lower stem has deeply lobed, stalkless, leaves in whorls of 3; upper leaves paired; basal leaves, if present, with 5-7 lobes, on long stalks
Habitat: wet meadows, prairies, shorelines
Similar Species: Wood Anemone, *A. quinquefolia*, has stalked leaves on stem

Angelica
Angelica atropurpurea L.

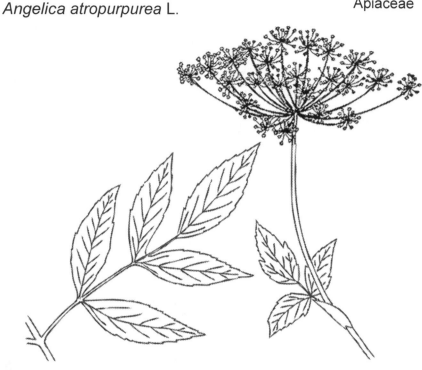

Plants: 60-240 cm tall; wide; robust; erect; perennial; smooth, thick, dark reddish stems

Flowers: 5 parted; greenish white to green; in a 10-20 cm wide umbel with a somewhat rounded top

Leaves: upper leaves with a swollen basal sheath; 3 main leaflets and each of which may be subdivided into 3 to 5 smaller leaflets

Habitat: wet meadows, streambanks, fens, swamps

Similar Species: Cow Parsnip, *Heracleum maximum*, has a larger, more rounded umbel; the three main leaflets are notched but not completely subdivided; stem is hollow

Groundnut
Apios americana Medik.

Status: FACW
Fabaceae

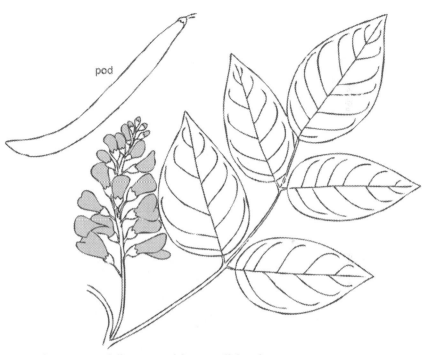

pod

Plant: vine, no tendrils; root with roundish tuber

Flower: purplish or brownish-purple; bilateral symmetry; in short racemes arising in leaf axils; distinctive scent

Leaves: alternate, compound; 5-7 leaflets, broad with sharp points; entire

Habitat: moist thickets

Similar Species: Hog Peanut, *Amphicarpea bracteata*, is a delicate vine with smaller pea-like flowers and three leaflets

Dragon's Mouth Orchid
Arethusa bulbosa L.

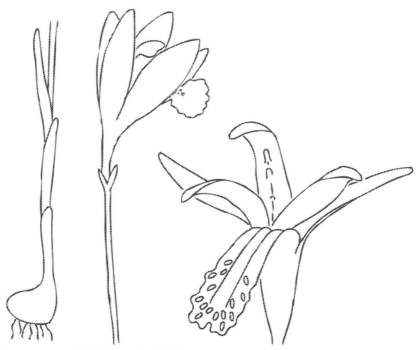

Plants: scape 10-25 cm; single flower at top
Flower: pink to magenta, 3-5 cm high; sepals erect, lanceolate, 20-55 mm x 3-9 mm; lateral petals curved forward, lanceolate, 23-49 mm x 4-10 mm; lower lip pinkish white, streaked with purple and yellow, curved or reflexed near middle, often slightly 3-lobed, with yellow lamellae and fleshy processes, 19-35 mm x 10-19 mm; flower subtended by two small bracts; flowers June-July
Fruit: erect capsule, ovoid, 15-25 mm
Leaves: single leaf on flowering stem, 2-4 mm wide; develops after flowering, nearly as long as flowering stem when full grown
Habitat: sphagnum bogs, coniferous and ericoid swamps, calcareous treed or open fens, moist acid sandy meadows
aka: Arethusa

Green Dragon
Arisaema dracontium (L.) Schott

Status: FACW
Araceae

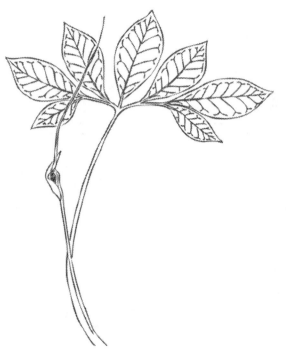

Plants: 30-120 cm tall; prennial; named after the elongate spadix that sticks out above the spathe, resembling a dragon's tongue
Flowers: white to pale yellow to pale green, tiny; crowded on a 10-20 cm long spadix inside of a hooded spathe
Fruits: orange to red berries
Leaves: one leaf forked to 5-15 unequal leaflets; palmately-shaped.
Habitat: swamps, streambanks

Jack in the Pulpit
Arisaema triphyllum (L.) Schott

Status: FACW-
Araceae

Leaves: Basal, usually 2. Petiole reddish-purple. Blade divided into 3 leaflets, each up to 30 cm long. Undersides of mature leaves whitish. Lateral leaflets assymetrical with the proximal (nearest the base) side having a rounded edge and the distal side having a straighter edge
Flower: On peduncle arising between leaves. Spathe wraps around itself at the base forming a greenish, often striped with purple, tube. Upper portion of the spathe curves forward and forms a hood over the spadix. Outside of hood purple or green with yellowish stripes along the veins; inside of hood purple with yellowish stripes along the veins. The hood is slightly ridged along veins. Cylindrical male or female flower (the spadix) is a reddish-purple and rod-like with a blunt end. During the summer the spathe disintegrates, exposing the spadix with its green, then red, berries
Habitat: Moist deciduous woods, floodplains, swamps
Similar Species: species is very variable but unique

Swamp Milkweed
Asclepias incarnata L.

flower

Plants: 60-180 cm tall; smooth, branched stems; large bright terminal blossoms; milky sap

Flowers: 20-30 flowers in cluster at top of stems; flower 5 parted, bright reddish purple; reflexed petals, each with a lighter accessory structure called the "hood" that expands upward and has inwardly curved "horns" longer than the hood itself; with a cinnamon scent; fruit a long pointed seedpod

Leaves: opposite; entire; lance-shaped, 15 cm long by 2 cm wide; leaf veins at acute angle to midrib

Habitat: wet meadows, swamps, prairies, marshes, bogs, shallow area along lakes and ponds, streambanks, ditches

Similar Species: distinguished from Common Milkweed, *A. syriaca*, and Prairie Milkweed, *A. sullivantii*, both of which have distinctly pink flowers and wider leaves

25

Yellow Screwstem
Bartonia virginica
(L.) Britton, Sterns & Poggenb.

Status: FACW
Gentianaceae

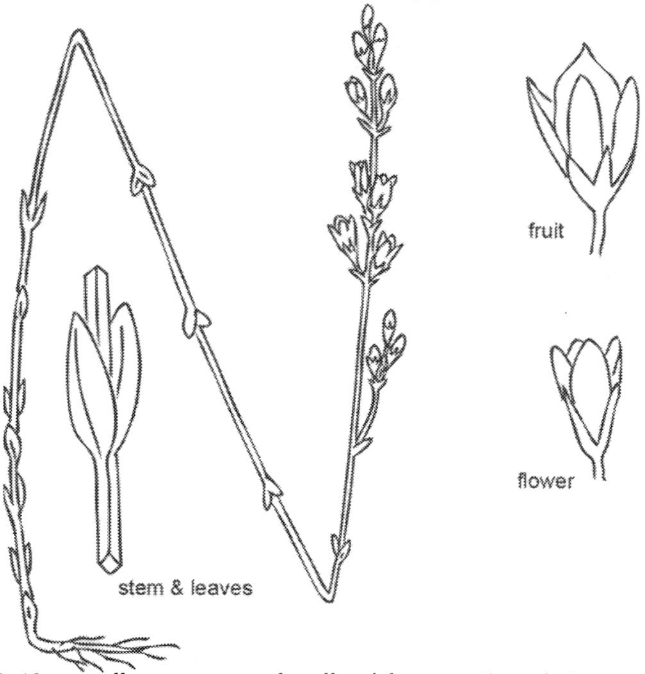

fruit

flower

stem & leaves

Plants: 10-40 cm tall; erect; annual, yellowish green 5-angled stems
Flowers: 4-parted; pale yellow to yellow; 4 oblong petals; 3-4 mm long
Leaves: opposite; scale like; tiny; up to 3 mm long; stalkless
Habitat: bogs, wet meadows, acidic ditches

Devil's Beggarticks
Bidens frodosa L.

seed

Plant: 30-120 cm high
Flower: yellowish; heads of tiny disk flowers surrounded by 5-9 green leafy bracts
Fruit: flat achenes with 2 long barbed awns
Leaves: opposite, compound; 3-5 lance-shaped toothed, but not lobed, leaflets
Habitat: damp ground, fields, waste places
Similar Species: Swamp Beggarticks, *B. connata*, is similar but the leaves are not divided into distinct leaflets although the bottom leaves may be lobed; Tall Beggarticks, *B. vulgata*, is very similar but has 10-20 bracts surrounding the flower head; Few-Bracted Beggarticks, *B. discoidea*, is also similar but has only 3-5 bracts; Leafy-Bracted Beggarticks, *B. tripartita*, has opposite simple or lobed leaves with wings along the stalks of the leaves and larger bracts around the flower head; Bur-Marigold, *B. cerna*, has yellow ray flowers; the closely related Water Marigold, *Megalodonta beckii*, is an aquatic form with finely divided submerged leaves

27

False Nettle
Boehmeria cylindria (L.) Sw.

Status: FACW+
Urticaceae

Plant: 40-100 cm tall; stem without stinging hairs
Flower: green-brown, tiny; plants in unbranched racemes growing from leaf axils of upper leaves
Leaves: alternate, coarsely toothed; ovate; 3 large nerves from base; spikes of flowers may be leafy at tips
Habitat: moist ground, shady places
Similar Species: Clearweed, *Pilea pumila*, also lacks nettles on the stem, but is smaller, has branching flower clusters, and succulent stems and leaves

Water Arum

Calla palustris L.

Plants: 12-25 cm tall; erect, perennial; from nodes on rhizomes
Flowers: white to yellowish white to green; tiny; inflorescence is a 2.5-5 cm long spadix partially surrounded by a 5 cm long white, oval, flat spathe
Fruits: round cluster of red berries
Leaves: basal; broadly heart-shaped with pointed tip; 5-15 cm long stalk
Habitat: bogs, fens

Grass Pink Orchid
Calopogon tuberosus
(L.) Britton, Sterns & Poggenb.

Status: FACW+
Orchidaceae

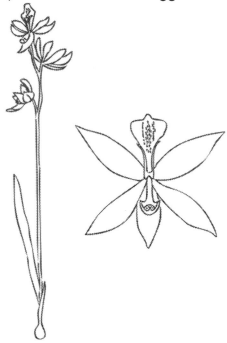

Plants: 30-65 cm tall; perennial

Flowers: 4-12 light to purplish pink flowers in terminal raceme; flower oriented so crested lip is at top; sepals 1.5-2.5 cm long and 5-13 mm wide; petals 1.5-2.5 cm long and 3-10 mm wide; upright lip with orange to yellow or whitish hairs

Leaves: single; basal; grass-like; 10-30 cm long, 3-16 mm wide

Habitat: bogs, fens, wet prairie

aka: *Calopogon pulchellus*

Yellow Marsh Marigold
Caltha palustris L.

Status: OBL
Ranunculaceae

Plant: 15-45 cm high; stems thick, hollow
Flower: yellow, 5 parted; 25-40 mm across, 5-9 waxy yellow sepals
Leaves: alternate, shallow toothed; glossy, rounded heart-shaped or kidney-shaped; long leafstalk
Habitat: swamps, wet meadows, brooksides
Similar Species: even without the flowers the thick leaves and the growth form are distinctive

Calypso Orchid
Calypso bulbosa (L.) Oakes

Status: FACW
Orchidaceae

Plants: 5-20 cm tall, erect; perennial; from bud-like corm
Flowers: usually solitary, at top of stem; sepals and petals pale pink to purplish pink, narrow lance-shaped, 1-2 cm long and 2.5-5.0 mm wide; oblong slipper-shaped lip, 1.5-2.5 cm long and 6-11 mm wide, yellow hairs near the opening, two small yellow projections at apex
Leaves: single; basal; 3-6 cm long, 2.0-4.5 cm wide
Habitat: bogs, swamps, lakeshores

Marsh Bellflower
Campanula aparinoides Pursh

Status: OBL
Campanulaceae

Plants: 15-90 cm tall; erect; stem thin and thread-like, 3 angled, with short hooking bristles; milky sap
Flowers: 5-parted; white to pale blue; 6-12 mm long; funnel-shaped; solitary; nodding
Fruits: capsule 1.2-2 mm long; three chambers
Leaves: alternate; lance-shaped, stalkless, long and narrow, up to 8 mm wide and 1.5-5 cm long; rough to touch; slightly toothed
Habitat: wet meadows, shorelines, streambanks, marshes

Pennslyvania Bittercress
Cardamine pensylvanica Muhl. ex Willd.

Status: OBL
Brassicaceae

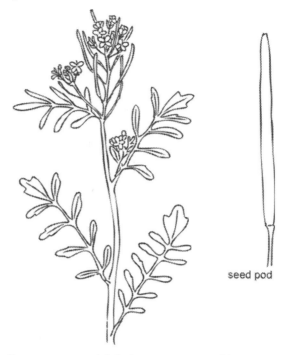

seed pod

Plants: 15-60 cm tall; erect; perennial; hairy stems toward base
Flowers: in loose clusters; 4-parted; white; tiny (~3 mm across)
Fruits: thin, long, erect pods
Leaves: basal and middle leaves elongate; deep, thin lobes apical lobe broader than lateral lobes
Habitat: swamps, streambanks, wet ground
Similar Species: Purple Cress, *C. douglassii*, is similar but has purple flowers and stem leaves that are stalkless and only slightly toothed; Cuckoo-Flower, *C. pratensis*, has leaves that are more finely divided and flowers that are pink to white and much larger (10-15 mm across)

White Turtlehead
Chelone glabra L.

Status: OBL
Scrophulariaceae

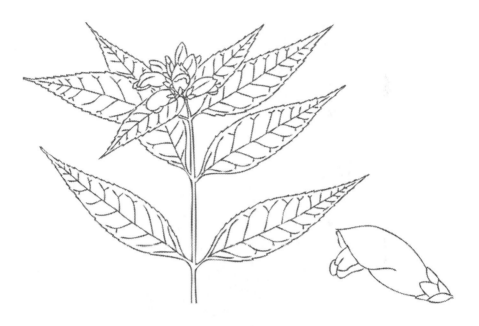

Plant: 30-100 cm high; erect; perennial, robust
Flower: flowers large (2-3 cm), clustered at top of stem; white sometimes tinged with pink; bilateral symmetry; two-lipped with upper lip extending over lower
Leaves: opposite, toothed, narrow lance-shaped, 7-15 cm long, without leaf-stalks
Habitat: wet ground, streambanks
Similar Species: easily distinguished when buds, flowers, or fruits present; plant and leaves somewhat similar to Common Vervain, *Verbena hastata*, which has leaves with longer stalks

American Golden Saxifrage
Chrysosplenium americanum
Schwein. ex. Hook.

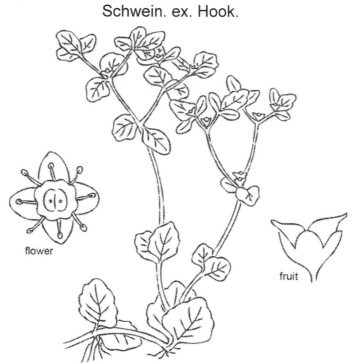

flower

fruit

Plants: 30-90 cm tall; erect; perennial; creeping stems
Flowers: solitary, at ends of branches; 4-parted; yellow to greenish, 4-5 mm across
Leaves: opposite; oval to round, less than 15 mm long
Habitat: muddy soil, usually in shade

Bulb-Bearing Water Hemlock
Cicuta bulbifera L.

Status: OBL
Apiaceae

Plant: 30-120 cm high; DEADLY POISONOUS
Flower: 5 parted, white; in umbels 3-5 cm wide
Leaves: alternate, twice compounded; leaflets very slender (less than 5 mm wide), sparsely toothed; may find bulblets in axils of upper leaves
Habitat: swamps, marshes
Similar Species: Water Hemlock, *C. maculata*, is similar but has wider leaflets, usually 5 mm or greater, and lacks the bulblets in axils of upper leaves

Broadleaf Enchanter's Nightshade
Circaea lutetiana L.

Status: FACU
Onagraceae

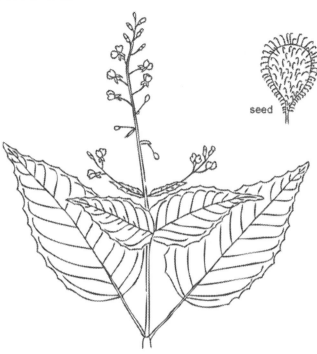

seed

Plants: 30-60 cm; erect; perennial; slender green stems; white hairs along central stem

Flowers: white; 2 deeply notched petals; small, 2 mm long and wide; arranged along stem at top of plant.

Leaves: opposite; ovate-cordate; toothed margins; up to 13 cm long and 10 cm wide

Habitat: swamps, upland forests

Similar Species: Small Enchanter's Nightshade, *C. alpina*, is smaller and has the flowers more clumped at the top, and leaves that are more heart-shaped

Swamp Thistle
Cirsium muticum Michx.

Status: OBL
Asteraceae

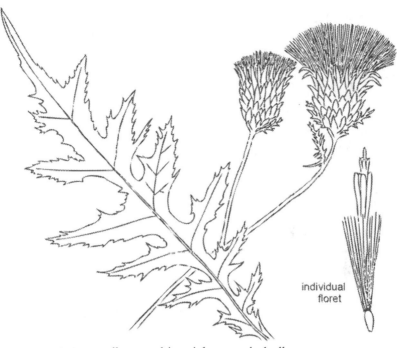

individual
floret

Plants: 60-210 cm tall; erect; biennial; smooth, hollow stems
Flowers: pink to purplish pink to purple disk flowers; 4 cm wide; flower heads surrounded by bracts with white, sticky, wooly (not spined) tips
Fruits: fluffy, feathery pappus
Leaves: deeply lobed lance-like to oblong; with sharp, stiff spines at tips of leaf lobes; up to 20 cm long and 8 cm wide
Habitat: wet meadows, streambanks, swamps
Similar Species: None of the following species have hollow stems: Pasture Thistle, *C. pumilum*, has thornless, but hairy, stems; Canada Thsitle, *C. arvense*, is very branched with small clusters of flowers; Tall Thistlte, *C. altissimum*, does not have deeply lobed leaves

Virgin's Bower
Clematis virginiana L.

Status: FAC
Ranunculaceae

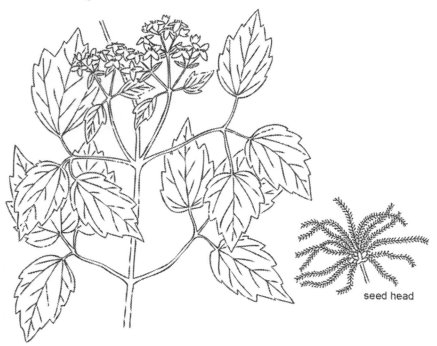

seed head

Plants: vines that climb by twining (no tendrils); lower parts woody and persistent, upper parts die back during winter

Flowers: several on a 2-6 cm stalk arising in the leaf axil; flowers with 4 white sepals 10-15 mm long, hairy (at least on the back)

Fruits: style becomes long (2-4 cm) and plume-like

Leaves: opposite-compound; 3 leaflets, 2-10 cm long, each coarsely toothed to 3-lobed; hairy beneath

Vines: climbs by twining; lower parts woody and persistent, upper parts ephemeral

Habitat: moist soils

Similar Species: Poison Ivy, *Toxicodendron radicans,* also has 3 leaflets but leaves are alternate

Purple Marshlocks
Comarum palustre L.

Status: OBL
Rosaceae

Plants: 20-60 cm tall; often sprawling; reddish brown stems from rhizomes
Flowers: 5-parted; 2.5 cm wide; petals half as long as sepals, both red to maroon
Fruits: smooth
Leaves: compound with 5-7 leaflets; long stalked; oblong to elliptical, sharply toothed; dark green on upper side; grayish on under side
Habitat: swamps, streambanks, marshes, edge of ponds, flooded meadows
aka: Marsh Cinquefoil, *Potentilla palustris*

Chinese Hemlockparsley
Conioselinum chinense
(L.) Britton, Sterns & Poggenb.

Status: FACW
Apiaceae

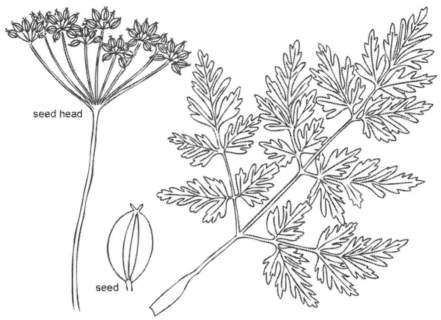

seed head

seed

Plants: 40-150 cm tall; erect; smooth, hollow, green stems
Flowers: 5-parted; white; flowering head compounded of dense umbellets at top of long stalk; 2.5-13.0 cm across
Fruits: splitting into 2 seeds
Leaves: compounded two to three times; leaflets deeply lobed, triangular-shaped
Habitat: bogs, fens, swamps, wet meadows
Similar Species: Queen Anne's Lace, *Dacuta carota*, is similar but the stem is covered with bristly hairs and there are deep, sharply lobed bracts under the umbel
Note: despite its name this species is native to North America

Threeleaf Goldthread
Coptis trifolia (L.) Salisb.

Status: FACW
Ranunculaceae

seed head

Plants: 5-15 cm tall; with yellow thread-like rhizome; flower early
Flowers: 5-7-parted; white; 1 cm across; star-shaped; 5-15 cm tall on a stalk; usually solitary; golden yellow club-shaped petals
Leaves: basal; compound with 3 fan-shaped, toothed, leaflets
Habitat: bogs, swamps
Similar Species: Barren Strawberry, *Waldsteinia fragarioides*, has similar leaves but the leaf stalks have small hairs; flowers (if present) are yellow and in small clusters

Bunchberry
Cornus canadensis L.

Status: FAC-
Cornaceae

Plants: 15 cm tall
Flowers: central cluster of small greenish to brownish flowers, surrounded by 4 larger white petal-like bracts
Fruits: bright red berries in terminal clusters; ~8 mm across
Leaves: opposite (appearing as whorls in a group of 4-6); shiny dark green above, paler below; very short petiole
Habitat: cold forested areas

Showy Lady's Slipper
Cypripedium reginae Walter

Plants: 30-90 cm tall; robust, 3-5 leaves per stem; arising from rhizome
Flowers: white petals and sepals; enlarged pouch-shaped lip with inrolled edges; lip white tinged with pink to red around opening
Leaves: ovate; 10-25 cm long and 4-16 cm wide; densely pubescent
Habitat: bogs, wet meadows, swamps
Similar Species: Leaves of Moccasin Flower, aka Pink Lady's Slipper, *C. acaule*, are basal. Ram's Head Lady's Slipper, *C.arietinum*, is much smaller and has a pointed lip

Swamp Loosestrife
Decodon verticillatus (L.) Elliot

Status: OBL
Lythraceae

Plants: 60-240 cm tall; unbranched; woody at the base with 4-6 ridges; often bent or reclining

Flowers: 5 petals; pink to purplish pink; bell-shaped, 1.0-1.5 cm acrosss petals; clustered in the axils of the uppermost leaves

Fruits: reddish to dark brown; 8 mm in diameter; spherical capsules

Leaves: opposite or sometimes whorled; elliptical to lance-shaped, pointed at tip, stalkless

Habitat: swamps, marshes, shallow water

Similar Species: Purple Loosestrife, *Lythrum salicaria,* and Winged Loosestrife, L. *alatum* have flowers in dense spikes at the top of the plant

herbaceous

Tall Flat-Topped White Aster
Doellingeria umbellata (P. Mill.) Nees

Status: FACW
Asteraceae

Plants: 60-150 cm tall; erect perennial arising from rhizome; light green to purplish red to yellowish brown centrol stem; central stem terminates in a somewhat flat-topped compound corymb of flowerheads

Flowers: each flower head (1.0-1.5 cm across) has 5-12 white to creamy ray florets surrounding 12-25 yellow disk florets which become purplish with age

Leaves: alternate; elliptical or lanceolate-shaped, 8-13 cm long, 12 to 25 mm wide, with rough edges, but not toothed; upper leaf surface is medium to dark green while the lower surface is pale green.

Habitat: wet meadows, borders of swamps

Similar Species: this species can be distinguished by the relatively flat top to the flower head and the relatively small white flowers (but see Smooth White Old Field Aster, *Symphyotrichum racemosum*)

aka: *Aster umbellatus*

Roundleaf Sundew
Drosera rotundifolia L.

Status: OBL
Droseraceae

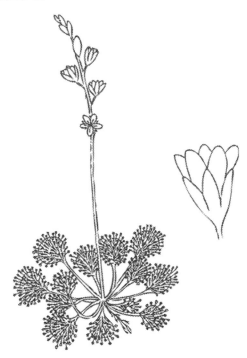

Plants: carnivorous; leaf folds around, and digests, insects that get caught on the sticky hairs

Flowers: white to pale pink; small, 1.25 cm across; 2-15 flowers on a 5-25 cm long stalk

Fruits: capsule with small light brown seeds

Leaves: from a basal rosette; round, 1.5-2.5 cm wide and long, on a long leaf stalk; upper surface of leaf blades and leaf stalks are covered with reddish, sticky glandular hairs; leaves lie flat on ground

Habitat: bogs, fens, swamps

Similar Species: Spatulateleaf Sundew, *D. intermedia*, has a more oval leaf and the leaf stalks lack the sticky hairs

Small Waterwort
Elatine minima (Nutt.) Fisch. & C.A. Mey.

Plants: small in size, to 5 cm tall; floating or in muddy areas; forms branched mats; shallow thread-like roots
Flowers: 2-4-parted; tiny
Fruits: capsule
Leaves: opposite; oblong-oval; 2.5-4.0 mm long with rounded tip; petiole absent
Habitat: shallow waters, mudflat

Black Crowberry
Empetrum nigrum L.

Status: FACW
Empetraceae

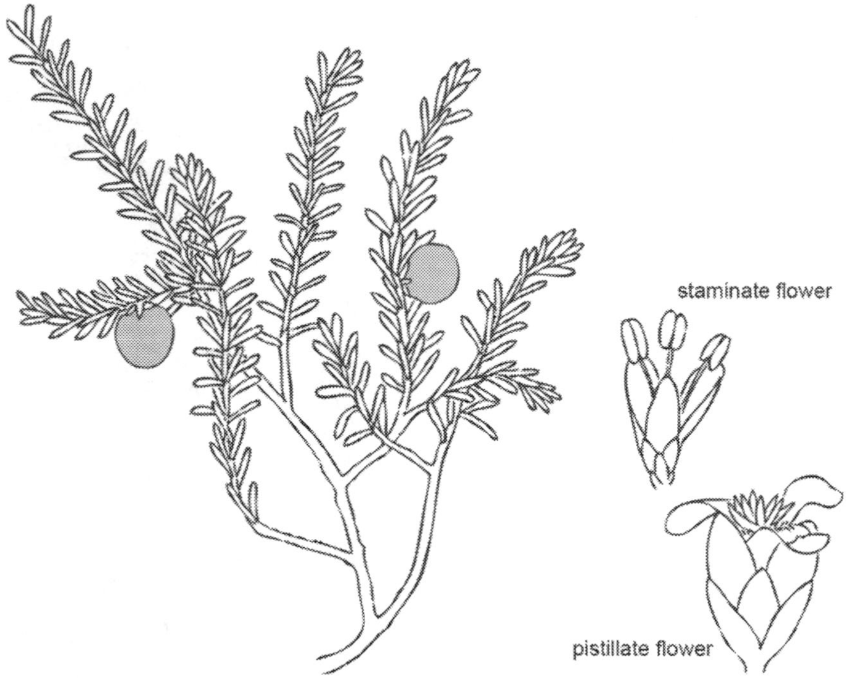
staminate flower
pistillate flower

Plants: up to 15 cm; low creeping evergreen, forms dense mats
Flowers: small solitary flowers in leaf axils; 3 petals; flowers either unisexual (males with 3 stamens) or bisexual (4 stamens); crimson to purple to purplish brown
Fruits: purplish then almost black berry; edible
Leaves: narrow, thin, needle-like, 3-8 mm long; dark green; in whorls of four
Habitat: open peaty soils of bogs, fens, swamps, and boreal forests

Purple-Leaved Willowherb
Epilobium coloratum Biehler

Plant: 30-90 cm; stem hairy, often purplish; stems erect from persistent basal rosettes

Flower: 4 parted, pink or sometimes white; small (less than 6 mm across); base of flower forms narrow stalk-like tube, petal notched at tip; in upper leaf axils, often nodding

Fruit: long, ascending seedpods; seeds with cinnamon-colored hairs

Leaves: alternate, with numerous small sharp teeth; narrow (1.0 cm wide or more), lance-shaped; dull gray-green, may be purplish in patches

Habitat: swamps and moist thickets, wet ground

Similar Species: Hornemann's Willowherb, *E. hornemannii,* is similar but has more sprawling stems that arise from elongated creeping stems on the ground surface; the other Willowherbs in our area have leaves without teeth

51

Downny Willowherb
Epilobium strictum Muhl. ex Spreng.

Status: OBL
Onagraceae

seed pod &
seeds

Plants: 30-60 cm tall; erect; unbranched; stems and leaves hairy; from rhizomes

Flowers: 4-parted; white to pale pink to pink; small; 5-8 mm across

Fruits: long, ascending seedpods; seeds with numerous dirty brown hairs

Leaves: alternate; narrow (less than 1.0 cm wide), lance-shaped; in-rolled downwards; hairy

Habitat: bogs, swamps

Similar Species: Marsh Willowherb, *E. palustre*, is similar but the upper surface of the leaves is usually hairless; the other Willowherbs in our area have toothed leaves

Sevenangle Pipewort
Eriocaulon aquaticum (Hill) Druce

Status: OBL
Eriocaulaceae

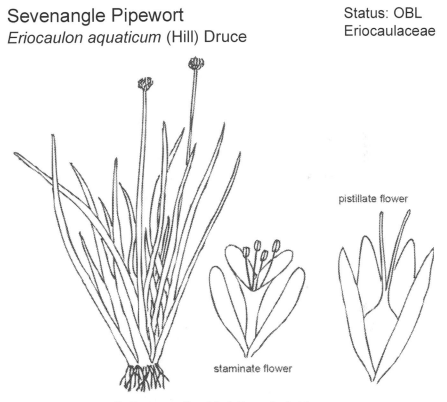

pistillate flower

staminate flower

Plants: 4-20 cm tall; flower stalk with 5-7 vertical ridges
Flowers: tiny flowers in a rounded flowerhead , 4-6 mm across, at top of stalk; unisexual, 3-parted; white
Fruits: pale brown to reddish brown; egg-shaped; ~0.5 mm long
Leaves: long, green, tapering (2-10 cm long, 2-5 mm wide), with 3-9 veins; from basal rosette
Habitat: shallow waters of shorelines, bogs
aka: *E. septangulare*

Spotted Joe-Pye Weed
Eupatorium maculatum L.

Plant: 60-200 cm tall; stem purple spotted or purple, not hollow
Flower: tiny, pinkish to purplish; in flat-topped cluster, each small head within cluster a composite of tiny individual flowers (immature flower clusters are more rounded; as they mature the cluster becomes flattened)
Leaves: whorled in 4's or 5's, toothed; single main vein; taper toward base; no scent when crushed
Habitat: wet meadows and thickets
Similar Species: other purplish-pinkish members of the genus have flower heads that are more rounded and have different leaf/stem characters; Hollow Joe-Pye Weed, *E. fistulosum*, is very similar but has a hollow stem
aka: *Eupatoriadelphia maculatum*

Boneset
Eupatorium perfoliatum L.

Plant: 60-150 cm high; stem hairy
Flower: tiny flowers in heads which are themselves arranged in a flat-topped cluster, grayish white
Leaves: opposite, toothed; veiny-wrinkled, taper-pointed leaves unite at bases around stem which pierces through the leaf tissue (perfoliate)
Habitat: low ground, thickets, swamp
Similar Species: the combination of the perfoliate leaves and the hairy stem are distinctive

Queen of the Meadow
Filipendula rubra (Hill) B.L. Robins.

Species: FACW
Rosaceae

Plant: 60-240 cm tall
Flower: 5 parted, dark pink; large divergently branching clusters at top of stem
Leaves: alternate, compound; leaves large, divided into 3-7 leaflets, these deeply lobed and sharply toothed; leafy stipule at base of leaf
Habitat: moist prairies, meadows, roadsides

Rough Bedstraw
Galium asperellum Michx.

Status: OBL
Rubiaceae

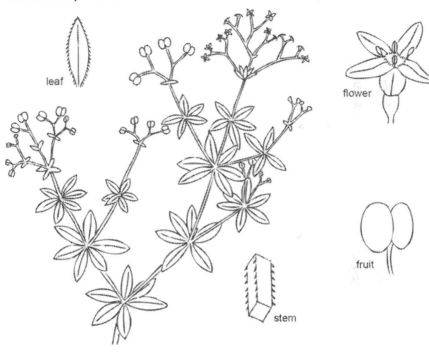

leaf

flower

stem

fruit

Plant: reclining on other plants; square stems covered with recurved prickles
Flower: 4 parted; white; flowers in numerous terminal clusters
Leaves: whorls of 6, sometimes 4 or 5; entire; leaves ~15 mm long, edged with recurved prickles
Habitat: damp woods, marshes, swamps
Similar Species: Cleavers, *G. aparine*, also recurved prickles on the stem, but has longer leaves (25-75 mm) and leaves mostly in whorls of 8

Marsh Bedstraw
Galium palustre L.

Plant: reclining on other plants; square stems, not covered with recurved prickles
Flower: 4 parted, white; flowers in loose terminal clusters
Leaves: whorled, entire; leaves 15-20 mm long, mostly in whorls of 4 to 6
Habitat: wet meadows, swamps
Similar Species: Northern Bedstraw, *G. boreale*, and Licorice Bedstraw, *G. circaezans* also have leaves mostly in whorls of 4, but in Northern Bedstraw the flowers are in tight clusters and in Licorice Bedstraw the leaves are broader; Threepetal Bedstraw, *G. triflorum*, has leaves mostly in whorls of 6 but the leaves are fragrant and the flowers have 3 parts born on long arching stalks

Creeping Snowberry
Gaultheria hispidula (L.) Muhl. ex Bigelow

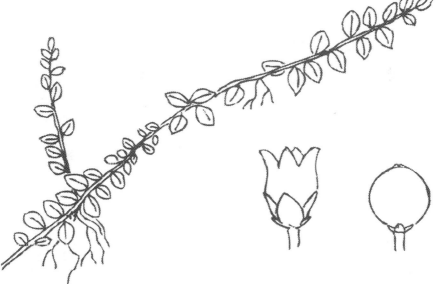

Plants: 2-4 cm tall; prostrate creeping stem with erect to sprawling branches (to 20 cm long); mat forming; evergreen

Flowers: 4-parted; white to pale pink, 3-5 mm long, bell-shaped; from leaf axils

Fruits: white berry with wintergreen flavor

Leaves: alternate; closely spaced, round to elliptical, leathery; small (3-7 mm long); shiny dark green with brown scale-like hairs; edges rolled under

Habitat: bogs, moist mossy woods

Closed Bottle Gentian
Gentiana andrewsii Grisb.

Plants: 30-60 cm tall; erect stem
Flowers: clustered atop stem and sometimes in axils of upper leaves; blue to violet; cylindrical bottle shape, 2.5 to 3.5 cm long; petals remain closed, joined to one another by a fringed white membrane that is just slightly longer than the petalsleaves
Leaves: opposite or sometimes whorled; lance-shaped, up to 15 cm long by 4 cm wide
Habitat: wet meadows
Similar Species: Bottle Gentian, *G. clausa*, is very similar but the whitish membrane joining the ends of the petals is shorter and can be seen only when the end of the flower is forced open

Greater Fringed Gentian
Gentianopsis crinita (Froel.) Ma

Status: OBL
Gentianaceae

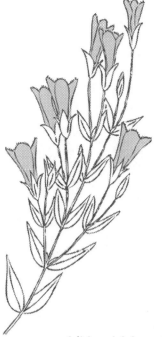

Plants: 30-90 cm tall; erect; annual/biennial; branched near the top
Flowers: solitary on long stalks; 4-parted; bright blue; funnel-shaped, 4-5 cm long; petals with long, delicate fringe on margin; open in the sun and close at night
Leaves: opposite; lance-shaped with wide rounded base; pointy tip; 1-2 cm wide; stalkless
Habitat: wet meadows, swamps, streambanks
Similar Species: the only Fringed Gentian, *Gentianopsis spp*, in the Adirondacks

Largeleaf Avens
Geum macrophyllum Willd.

Plants: 30-90 cm tall; stem erect, hairy

Flowers: yellow; in asymmetrical open cluster; 5 ovate to rounded petals, 4-6 mm long; petals longer than sepal, pointed reflexed sepals

Fruits: yellow-green, hairy; elliptical; 3 mm long

Leaves: Basal Leaves: long stemmed compound leaves: end leaflet large, triangular-ovate to heart-shaped, small-toothed to deeply lobed; side leaflets much smaller; up to 45 cm long; Stem Leaves: alternate leaves with 3 pointed lobes, 11 cm long and 9 cm wide; both types of leaves coarsely toothed and hairy

Habitat: swamps, wet meadows, woods

Similar Species: in the basal leaves of Yellow Avens, *G. aleppium,* the end leaflets is more triangular and deeply lobed; Rough Avens, *G. lacinatum,* is similar to Yellow Avens but the petals are much shorter than the sepals; Agrimonies, *Agrimonia spp,* have leaves resembling the Avens but the flowers are born on a slender raceme

Purple Avens
Geum rivale L.

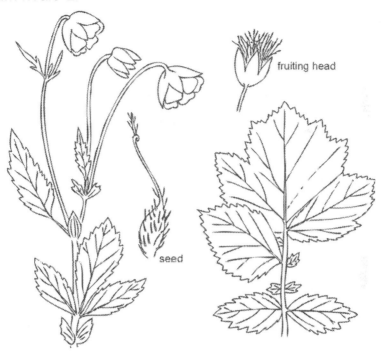

fruiting head

seed

Plant: 30-60 cm tall
Flower: 5 parted, purplish; ~12 mm long; nodding, globular; purplish sepals, yellow petals about as long as sepals
Fruit: tips of developing seeds form long hooked appendages
Leaves: alternate, compound; segments toothed and (end segment) 3 lobed; end segment broader than others; often with smaller leaflets interspersed along main axis of leaf
Habitat: swamps, wet meadows
Similar Species: other Avens, *Geum spp*, and Agrimonies, *Agrimonia spp*, have similar leaves, but the flowers are very different

Clammy Hedgehyssop
Gratiola neglecta Torr.

Status: OBL
Scrophulariaceae

Plants: 8-30 cm tall; erect; annual; light green stems with pubescent
Flowers: 5-parted; 2 lipped: upper lip forms 2 lobes, lower lip forms 3 lobes; white to yellowish; tubular; 7-12 mm long
Fruits: rounded capsule; 4-5 mm long
Leaves: opposite; lance shape to oblong, pointed at the tip, tapered at the base, slightly toothed; 1-5 cm long and 2-15 mm wide; light to medium green; hairless; may appear to have tiny bump-like glands
Habitat: wet muddy places: shorelines, edge of ponds and lakes, swamps, marshes, wet meadows
Similar Species: Golden Hedgehyssop, *G. aurea*, has leaves without teeth and bright yellow flowers

Common Mare's Tail
Hippuris vulgaris L.

Status: OBL
Hippuridaceae

Plants: mostly to 60 cm long; soft, unbranched, hollow stems; either underwater or emergent: somewhat erect when emergent, creeping when submersed

Flowers: small; in axils of upper leaves

Fruits: tiny, 2 mm long, nut-like

Leaves: whorled, 6-12; stalkless, smooth margins; submersed leaves are pale green and up to 5 cm long and 3 mm wide; aerial leaves usually degenerate quickly, are dark green and up to 3 cm long

Habitat: shallow areas of streams and lakes

Marsh-Pennywort
Hydrocotyle americana L.

herbaceous
Status: OBL
Apiaceae

Plant: creeping; flowers and runners in leaf axils
Flower: 5 parted, white; 1-5 tiny flowers and runners in leaf axils
Leaves: alternate, shallow scalloped teeth; rounded kidney-shape
Habitat: open muddy places
Similar Species: Moneywort, *Lysimachia nummularia*, has somewhat similar shaped leaves and growth habit, but the leaves are opposite

Pale St. John's Wort
Hypericum ellipticum Hook

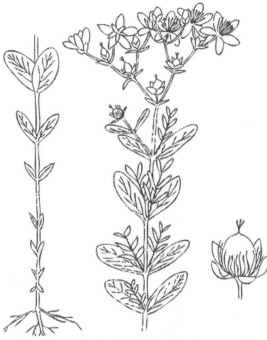

Plants: 20-40 cm tall; unbranched, slightly 4-sided stem arising from rhizome

Flowers: yellow, 5 parted; 12-18 mm across; styles fused to form a single beak, stamens 20 or more, petals separate, without dark dots

Fruits: rounded with short beak

Leaves: opposite, 1-3 cm long, pointed, no teeth; green dotted with translucent or black dots

Habitat: marshes, shores

Similar Species: except for the Common St John's Wort, *Hypericum perforatum* (see below), all of the other herbaceous St John's Worts in the Adirondacks have fewer than 20 stamens

Dwarf St John's Wort

Hypericum mutilum L.

Status: FACW
Clusiaceae

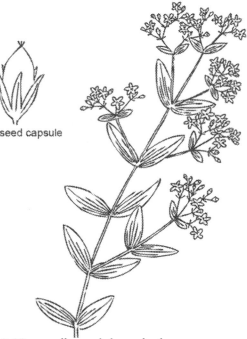

seed capsule

Plants: erect, 10-80 cm tall; much branched
Flowers: yellow; 5-parted, small (< 4 mm across; petals 1.5-2.5 mm); stamens few than 20; sepals are pointed and about the same length as the fruit (if present); bracts below flower clusters are small (1-4 mm) and awl-shaped
Fruits: capsule, green, 2.0-3.5 mm long
Leaves: opposite; lance-shaped to elliptical, 1-4 cm long, blunt to rounded at base, sessile, blunt to somewhat pointed at tip
Habitat: wet soil
Similar Species: very similar to terrestrial form of Northern St John's Wort, *H. boreale* - the most reliable difference seems to be that there the bracts below the flower clusters are wider and more leaflike, also the sepals are blunt and much shorter than the fruit; Canadian St John's Wort, *H. canadensis*, and Larger St John's Wort, *H. majus*, both have narrower leaves, with the former being 5-10 x longer than wide, tapering toward the base and the latter being 3-4 x longer than wide, rounded at the base

Common St John's Wort

Hypericum perforatum L.

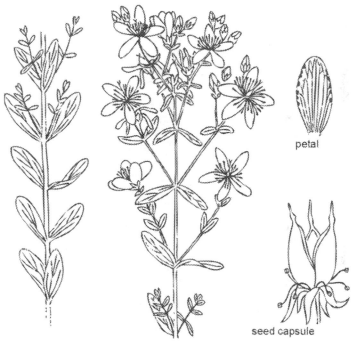

petal

seed capsule

Plant: 30-75 cm tall; stem branched
Flower: yellow, 5 parted; 18-25 mm across, stamens 20 or more, petals
separate, with dark dots along margins
Leaves: opposite, simple; less than 4 cm long, green with translucent dots
Habitat: roadsides, fields
Similar Species: Pale St John's Wort, *H. ellipticum*, the only other species in
the Adirondacks with at least 20 stamens has an unbranched stem and the
edges of the petals lack dark dots; Shrubby St John's Wort, *H. prolificum*, can
also be found in wetland habitats; superficially somewhat similar to Swamp
Candles, *Lysiimachia terrestris*

Jewelweed
Impatiens capensis Meerb.

Status: FACW
Balsaminaceae

Plant: 60-150 cm tall; stem succulent, exudes juice when broken
Flower: bilateral, longer than broad; orange with red-brown spots; pendant; spur at rear bends forward until parallel with flower
Fruit: elongate pod that springs open explosively when ripe
Leaves: alternate, coarsely toothed; egg-shaped, succulent
Habitat: wet, shady places
Similar Species: the flowers of Pale Touch-Me-Not, I. pallida, are yellow and have fewer spots; the spur at the rear bends only to a right angle with the flower; when flowers and fruits are absent the leaves could be confused with Clearweed, *Pilea pumila*

Yellow Flag
Iris pseudacorus L.

Status: OBL
Iridaceae

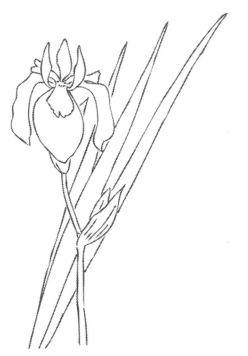

Plant: 30-90 cm tall, usually in large patches
Flower: yellow, 3 parted; up to 10 cm wide
Leaves: alternate, simple; although the leaves appear at first to be basal, closer inspection shows them to be alternate; parallel veined
Habitat: marshes, banks of streams
Similar Species: Yellow Flag, *I. pseudacorus*, is the only *Iris* with yellow flowers. After flowering it can be distinguished from Blue Flag, *I. versicolor*, its larger leaves (up to 130 cm long, 3.7 cm wide) and by its tendency to grow in large patches rather than in smaller discrete clumps

Blue Flag

Iris versicolor L.

seed pod

Plant: 30-90 cm tall, usually in small discrete clumps
Flower: blue, with white and yellow areas with dark veining and spots on the "fall" sepals, 3 parted; up to 10 cm wide
Leaves: alternate, simple; although the leaves appear at first to be basal, closer inspection shows them to be alternate; parallel veined
Habitat: marshes, wet meadows
Similar Species: Blue Flag, *I. versicolor,* is our only *Iris* with blue flowers. After flowering it can be distinguished from Yellow Flag, *I. pseudacorus,* by its shorter, narrower leaves and by its tendency to grow in discrete patches rather than in large patches

American Water-Willow
Justicia americana (L.) Vahl

Status: OBL
Acanthaceae

Plants: 40-100 cm tall; erect; with rhizomes; 4-angled stems
Flowers: clustered on top of stalks rising from leaf axils; white to pale purple with purple spots; lip of upper petal forms hood; elongate lower lip 3 lobed
Fruits: 4-seeded capsule
Leaves: opposite; slender; 8-16 cm by 0.8-2.5 cm; sessile or very short petiolate
Habitat: marshes, edge of lakes and ponds, streambanks

Wood Nettle

Laportea canadensis (L.) Weddell

Plant: 30-120 cm tall; stem covered with stinging hairs
Flower: greenish, tiny; in loose branching clusters arising from leaf axils
Leaves: opposite, coarsely toothed; egg-shaped; 7-15 cm long
Habitat: wet or moist woods
Similar Species: Stinging Nettle, *Urtica dioica*, also has stinging hairs on the stem, but the leaves are opposite

Marsh Pea
Lathyrus palustris L.

Status: FACW+
Fabaceae

Plants: slender trailing or climbing vines, stems winged or wingless
Flowers: long stalk with cluster of 2-6; pea-like, bilateral symmetry, 12-20 mm long, reddish purple
Fruits: flat, pea-like
Leaves: leaves compound, usually 6 leaflets (sometimes 4, 8, or even 10); leaflets elongated ovals, 2-8 cm long by 3-20 mm wide; tendrils at the tips of leaves; stipules at base of leaves asymmetrical - lobes unequal
Habitat: moist meadows, marshes, wet woods, shores
Similar Species: Beach Pea, *L. japonicus* aka *L. maritimus*, is a more robust species with more succulent leaves, it has larger stipules with equal lobes; Pale Vetchling, *L. ochroleucus*, has yellowish white flowers
aka: Marsh Vetchling

Canada Lily
Lilium canadense L.

herbaceous

Status: FAC+
Liliaceae

Plant: 60-150 cm tall,
Flower: 6 parted, orange, sometimes yellow or red; petals often brown-spotted; flowers large (5-10 cm across), nodding, in clusters on top of stem
Leaves: whorled, entire; lance-shaped; parallel veins, rough to touch on underside of leaves
Habitat: moist meadows, ditches, streamsides

Twinflower

Linnaea borealis L.

Status: FAC
Caprifoliaceae

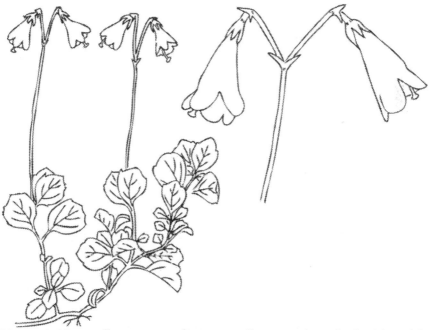

Plants: 8-15 cm tall; evergreen; form sprawling mats; branched; with aerial stems rising from stolon; stems become woody with age
Flowers: nodding, bell-shaped, ~12 mm long; 5 pink to pinkish white petals, 5 green sepals; in pairs on "Y" shaped stalk
Fruits: small one-seeded capsule
Leaves: opposite; small (~2 cm), egg-shaped to rounded; glossy; persisting for 2 years
Habitat: swamps, boreal forests

Bog Twayblade
Liparis loeseli (L.)

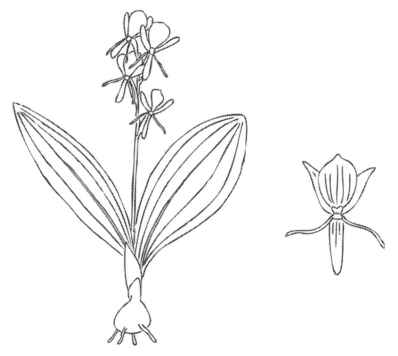

Plants: 10-25 cm tall; single hollow stem; with pseudobulb
Flowers: 6-parted; yellowish green; form a loose cluster of 2-15 flowers along stalk; sepals oblanceolate, petals linear, labellum oblong, darker in color
Leaves: basal, ovate to elliptical to oblong; 4-10 cm long and 2-5 cm wide; slightly folded with definite keel; yellowish green
Habitat: damp woods

Heartleaf Twayblade Orchid
Listera cordata (L.) R. Br.

Status: FACW+
Orchidaceae

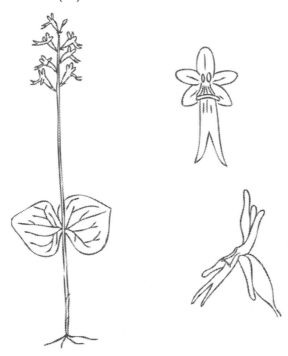

Plants: 10-25 cm tall; erect; hairy stems

Flowers: 6-21 flowers on tall stalk; greenish to reddish purple; 3 sepals and 2 petals in star-like arrangement; deeply-forked lip (notched about one half way up) resembles snake's tongue

Leaves: opposite, midway up stem; heart-shaped, 22-37 mm long and 22-35 cm wide; hairless and stalkless

Habitat: bogs, evergreen swamps, moist woods

Similar Species: Broad-Leaved Twayblade, *L. convallarioides*, has greenish yellow flowers with a lip that has only a shallow notch; Auricled Twayblade, *L. auriculata*, has pale greenish to pale bluish flowers with a lip that is notched 1/5 to 1/3 of the way up and the base of the lip has small wings at the base of the lip that curl up around and clasp the base of the column

American Shoreweed
Littorella uniflora (L.) Asch.

pistillate flower

staminate flower

Plants: perennial; emergent or submersed

Flowers: only in emergent plants, unisexual; 4-parted; staminate flowers white to greenish; tiny

Leaves: form basal rosette, tapering, dark green

Habitat: in wet soft sandy substrate of marshes, edges of lakes and ponds, shorelines, streambanks

Similar Species: Submersed plants resemble Quillworts, *Isoetes spp*, and Water Lobelia, *Lobelia dortmanna*

Cardinal Flower
Lobelia cardinalis L.

Status: FACW+
Campanulaceae

Plant: 60-120 cm tall
Flower: bilateral symmetry, bright red, large (25-35 mm long); lower lip split in 3, upper lip split in 2 with long stamens projecting through split
Leaves: alternate, toothed
Habitat: streambanks, wet meadows

Water Lobelia
Lobelia dortmanna L.

Status: OBL
Campanulaceae

Plants: 10-50 cm tall; erect; perennial; unbranched; smooth hollow stems
Flowers: loose cluster of flowers along a hollow stalk arising from the center of basal rosette; 5-parted; white to pale blue; tubular-shaped, 12-15 mm long, upper two petals curled upward, lower three petals curled downward
Fruits: capsule; 5-10 mm long and 3-5 mm wide.
Leaves: basal rosette; long (up to 8 cm), round, narrow, hollow, usually submerged; often curved outward
Habitat: marshes, edge of lakes and ponds, shorelines, streambanks
Similar Species: Brook Lobelia, *L. kalmii*, is similar but has narrow leaves on the stem and spatulate basal leaves; Indian Tobacco, *L. inflata*, has sessile flowers with inflated bases in the upper leaf axils, leaves ovate, hairy; Great Blue Lobela, *L. siphlitica*, is much larger and has larger flowers which are streaked with white on the lowe petal

82

Water-Purslane, Marsh Seedbox
Ludwigia palustris (L.) Elliot

Status: OBL
Onagraceae

Plants: 8-30 cm long; perennial; creeping or floating; light green to red stems; upper stems and leaves usually are above water
Flowers: small single flowers in leaf axils, only on emergent stems; small, four parted, green
Fruits: greenish 4-chambered capsule
Leaves: opposite; green to reddish green to purplish green; lance-shaped to oval, pointed tip, 10-25 mm long, smooth margin
Habitat: marshes, swamps, streambanks, shorelines

Cut-Leaf Water Horehound
Lycopus americanus Muhl. ex W. Bart.

Plant: 15-60 cm, stems square
Flower: bilateral, white; lower lip has 3 lobes, the upper 2; flowers tiny (~4-5 mm) in dense clusters in axils of leaves
Leaves: opposite, toothed; teeth of basal leaves deeply cut, almost lobes, upper leaves teeth not as deeply cut; not strongly mint-scented
Habitat: wet places
Similar Species: Water Horehound, *L. virginicus*, and Northern Water Horehound, *L. uniflorus*, do not have teeth on basal leaves as deeply cut; similar to Wild Mint, *Mentha arvensis*, but the leaves do not have a strong mint scent

Northern Water Horehound
Lycopus uniflorus Michx.

Plant: 15-60 cm, stems square
Flower: bilateral, white; lower lip has 3 lobes, the upper 2; flowers tiny (~4-5 mm) in dense clusters in axils of leaves
Leaves: opposite, toothed; teeth of basal leaves coarsely cut but not almost lobed; leaves light green; not mint-scented
Habitat: wet places
Similar Species: Water Horehound, *L. virginicus*, has very similar leaves but they are dark-green or purplish and note (use a hand lens) that the lower lip of its flower has 2 lobes, not 3; Cut-Leaf Water Horehound, *L. americanus*, has basal leaves which are cut so deeply that they are almost compound; somewhat similar to Wild Mint, *Mentha arvensis*, but the leaves do not have a strong mint scent

Fringed Loosestrife
Lysimachia ciliata L.

Status: FACW-
Primulaceae

Plant: 30-120 cm tall

Flower: growing singly nodding on long stalks; yellow, 5 parted, 12-25 mm across; petals roundish, toothed or pointed on end, slightly joined at base, unspotted

Leaves: opposite, entire; egg-shaped, on long stalks fringed with hairs

Habitat: swamps, wet thickets, shores

Similar Species: Whorled Loosestrife, *L. quadrifolia*, has flowers with red dots around the center and leaves lacking leafstalks in whorls of 4 (sometimes 3 or 5)

Note: all of these species of *Lysimachia*, except *nummularia*, have been given the common name of Swamp Candles by one field guide or another

Moneywort
Lysimachia nummularia L.

Status: OBL
Primulaceae

Plant: creeping, trailing plant, 15-60 cm long
Flower: 5 parted, yellow; ~20 mm across, long stalked arising in leaf axils, often paired
Leaves: alternate, entire; small, roundish, shining; leafstalks short to lacking
Habitat: moist open ground
Similar Species: Marsh Pennywort, *Hydrocotyl americana*, has a similar growth form, but has alternate leaves
aka: Creeping Jenny

Swamp Candles

Lysimachia terrestris (L.) B. S. P.

Plants: 30-100 cm tall
Flowers: terminal raceme; yellow; 5 parted, 6-13 mm across; petal rounded but longer than wide, slightly joined at base, red dots around center
Leaves: opposite; not toothed; lance-shaped, 25-75 mm long
Habitat: swamps, marshes, wet shores
Similar Species: distinguished from other yellow colored Loosestrifes, *Lysimachia spp*, by flowers growing in terminal raceme

Swamp Loosestrife
Lysimachia thrysiflora L.

Plant: 50-100 cm tall
Flower: densely packed heads, growing on short raceme, arising in pairs from leaf axils; yellow, 5 parted, ~5 mm across; petals, 5 sometimes 6, somewhat longer than wide, slightly joined at base
Leaves: opposite, entire; lance shaped, without leafstalk, 5-10 cm long
Habitat: cold swamps
Similar Species: only Loosestrife, *Lysimachia spp*, with clusters of flowers growing in racemes from the axils of the leaves

Purple Loosestrife
Lythrum salicaria L.

Plant: 60-120 cm tall
Flower: 6 parted (sometimes 4 or 5), 12-18 mm wide, purple; multiple blooms in upper leaf axils forming a dense spike at top of plant
Leaves: opposite (sometimes in whorls of 3); lance-shaped, widest at the base, almost clasping stem, entire, downy
Habitat: swamps and wet meadows, sometimes growing in large colonies
Similar Species: Winged Loosestrife, *Lythrum alatum*, has single flowers in the upper leaf axils and a square stem

Canada Mayflower
Maianthemum canadense Desf.

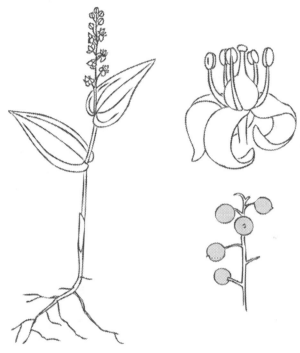

Plant: 7-15 cm
Flower: white, 4 parted; flowers in a small raceme
Fruit: berries white, then turning pale red
Leaves: alternate, entire; usually 2, sometimes 3; deeply cleft heart-shape base; parallel veins
Habitat: woods
Similar Species: Starry False Solomon's Seal, *M. stellata*, and False Solomon's Seal, *M. racemosa*, have more leaves and 6 parted flowers

91

Starry False Solomon's Seal
Maianthemum stellata (L.) Link

Plant: 30-60 cm high
Flower: white, 6 parted
Fruit: berries striped with black when mature
Leaves: alternate, entire; leaves clasp stem
Habitat: moist open places
Similar Species: the larger number of leaves distinguish this from Canada Mayflower, *M. canadense* (also, it has 4 parted flowers); False Solomon's Seal, *M. racemosa*, has somewhat smaller flowers on branched racemes and berries which are white then red, its leaves do not clasp the stem as in *M. stellata* nor are they as close together on the stalk
aka: *Smilacina stellata*

Wild Mint
Mentha arvensis L.

Plants: 15-50 cm tall; square stem, more or less hairy
Flowers: tiny clusters in leaf axils; bilateral symmetry, pale violet or sometimes white
Leaves: opposite, toothed; 7-15 cm long, egg-shaped to oblong; smells strongly of mint
Habitat: moist or wet open spaces, shores
Similar Species: the Water Horehound, *Lycopus spp*, do not have leaves with a strong mint scent
aka: *M. canadensis*

Buckbean
Menyanthes trifoliata L.

Plants: emergent with leaves and flower above water; arising from creeping rootstalk
Flowers: clustered at top of stalk; 5-parted, star-shaped, ~15 mm across; white to pink, petals are fringed with long white hairs
Fruits: ellipsoid capsule, 8-10 mm long
Leaves: alternate; compound with 3 leaflets; leaflets oblong to elliptical, 5-10 cm long and 2.5 to 5 cm wide, toothless, not hairy
Habitat: shallow water of bogs and ponds

Alleghany Monkey Flower
Mimulus ringens L.

Status: OBL
Scrophulariaceae

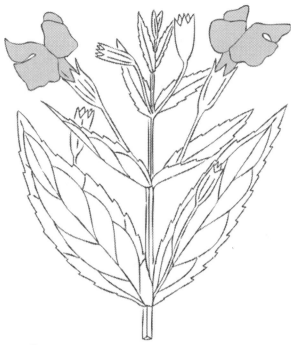

Plant: 30-90 cm tall; stem square
Flower: bilateral, blue with yellow throat; ~25 mm long; lower lip with 3 spreading lobes; flowers in upper leaf axils on long stalks
Leaves: opposite, toothed; without leaf stalks
Habitat: swamps, streamsides, wet ditches
Similar Species: Winged Monkey Flower, *M. alata,* has squarer stems, the angles of which often have wing-like extensions, leaves are on short stalks, flowers on short stalks

Naked Miterwort
Mitella nuda L.

Status: FACW-
Saxifragaceae

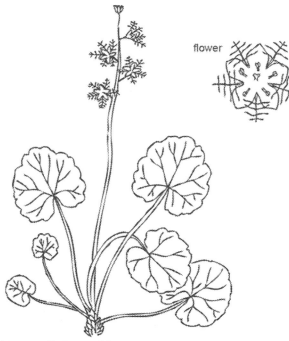

flower

Plants: 15-30 cm tall; from rhizome
Flowers: loosely cluster along a central stalk arising from the center of basal leaves; 6-8 mm across; 5 parted, fringed petals resemble snowflake, white to greenish
Fruits: capsules, 2-3 mm long
Leaves: basal leaves; rounded to heart-shaped; long stalked; round-toothed; 2-5 cm wide; with stiff hairs
Habitat: cool woods, swamps, bogs
Similar Species: Twoleaf Miterwort, *M. diphylla*, has basal leaves with sharper lobes and teeth and a pair of opposite leaves on the stem

True Forget-Me-Not
Myosotis scorpioides L.

Plant: 15-60 cm; sprawling to weakly erect; stems downy, angled
Flower: 5 petals, blue with yellow center; 6-10 mm across; lobes of calyx are one-quarter to one-third its total length; flowers on two diverging branches which uncoil as flowers bloom
Leaves: alternate, sessile; entire, 25-75 mm long, oblong to lance shaped, hairy
Habitat: springs, brooksides, muddy shores
Similar Species: Bay Forget-Me-Not, *M. laxa*, has smaller flowers (2-6 mm) and lobes of its calyx are about 1/2 its total length; Spring Forget-Me-Not (aka Early Scorpion Grass), *M. verna*, has inconspicuous small white flowers

Bog Aster
Oclemena nemoralis (Aiton) Greene

Plants: 20-70 cm tall; erect; perennial
Flowers: solitary at top of stem; 12-25 ray florets and 20-35 disc florets, 2.5-4 cm across; pink to pale purple; floral bracts very narrow, purple tinged
Leaves: alternate; lance-shaped, to 6 cm long, tapered at both ends, rough margins but toothless; stiff, with inrolled margins
Habitat: bogs
Similar Species: most other Asters have multiple flower heads
aka: *Aster nemoralis*

Scheinitz's Ragwort
Packera schweinitziana Weber & Love

Status: FACW
Asteraceae

Plants: 40-70 cm tall; perennial, from rhizome
Flowers: clusters of flowers at ends of stems, on long stalks; individual composites 5-12 mm wide, 8-15 yellow ray florets surrounding disk florets
Leaves: small basal leaves, 3-7 cm long by 1-2 cm wide, toothed margins; stem leaves toothed distally and deeply lobed
Habitat: marshes, swamps, ditches
Similar Species: Roundleaf Ragwort, *P. obovata*, has rounded basal leaves
aka: *Senecio schweinitziana*

Green Arrow Arum
Peltandra virginica (L.) Schott

Plants: 30-60 cm; erect; perennial

Flowers: small pale yellow to pale green flowers located on slender spathe wrapped in an elongated (10-18 cm) green spadix

Fruits: cluster of green berries

Leaves: basal, long stalked; shapes highly variable, most commonly arrow-shaped 10-40 cm; upper side green to purplish green, whitish under side with three prominent veins

Habitat: marshes, swamps, shallow area of ponds and slow-moving streams

Clearweed
Pilea pumila (L.) A. Gray

Plants: 10-30 cm tall; stems lacks stinging hairs, fleshy and translucent; often form dense colonies

Flowers: tiny, greenish, ~1 mm wide; in short drooping clusters on stalks in the leaf axils

Fruits: pale green usually with slightly raised purplish spots, 1.3-2.0 mm long

Leaves: opposite, egg-shaped with rounded or wedge-shaped base, 3-12 cm long with, 11-17 blunt teeth; long leafstalk; lacks stinging hairs

Habitat: moist shady places, shaded streambanks

Similar Species: Stinging Nettle, *Urtica dioica*, and Wood Nettle, *Laportea canadensis*, both have stinging hairs on the stems and leaves; False Nettle, *Boehmeria cylindrica*, is larger and has flowers in more compact clusters strung like beads along the flower stalk

101

White Fringed Orchid
Platanthera blephariglottis (Willd.) Lindl.

Status: OBL
Orchidaceae

Plants: 30-90 cm; perennial; slender
Flowers: white; 2.5-4.0 cm long; lip long (1-2 cm), tongue-shaped (not lobed), finely fringed; elongate spur
Leaves: green; lower ones linear to lance-shaped, to 20 cm long by 2 cm wide; upper leaves much reduced
Habitat: wet meadows, bogs, fens
Similar Species: There are 11 other species of *Platanthera* found in the Adirondacks, all of which might be encountered in wetlands or moist areas
aka: these *Platanthera spp* are sometimes placed in the genus *Habernia*

Rose Pogonia Orchid
Pogonia ophioglossoides (L.) Ker Gawl.

Status: OBL
Orchidaceae

Plant: 10- 30 cm tall; stem hairless, purplish at the base
Flower: solitary; pale pink to pink; 6 parted, up to 5 cm long; lower petal
has fringed lip with yellow hairs and red to dark purple veins.
Leaves: usually one per stem sometimes two (if so, then alternate); sessile,
lance-shaped to elliptical, pointed; base of leaf partly wraps around the
stem; up to 7.0 cm long by 2.5 cm wide
Habitat: evergreen shrub swamps, wet meadows, bogs
Similar Species: In Dragon's Mouth Orchid, *Arethusa bulbosa*, the leaf
develops after flowering; in Calypso Orchid, *Calypso bulbosa*, the sepals and
petals are much smaller than the lip

Smartweed
Polygonum amphibium L.

Status: OBL
Polygonaceae

Plants: has both a terrestrial variety (shown above) and an aquatic variety; the form of the species is plastic depending on water levels; stems and sheaths hairy

Flowers: 5 parted, rose; small flowers in blunt spike about 12 mm wide and 25 mm long (aboout half as wide as long)

Leaves: leaves rise above the water or lie flat on the water surface, depending on the variety; alternate, entire; short leafstalk, in this genus the base of the leaf stalk swells at the node where it joins the stem and wraps around the stem, this sheath is hairy in this species

Habitat: shallow water in swamps, marshes, ditches

Marshpepper Smartweed
Polygonum hydropiper L.

Plant: 10- 60 cm tall; stem reddish; sheaths of joints usually not fringed; from taproot

Flower: 4 parted (sometimes 5), greenish-white; slender spikes 25-75 mm long, drooping at tips

Leaves: alternate, entire; wavy margins; extremely acrid-peppery to taste; sheaths at nodes (see *page 104*) wrap completely around stem and overlap in front

Habitat: moist soil, shores

Similar Species: Swamp Smartweed, *P. hydropiperoides*, is similar, although the flowers are 5 parted and are usually pink, the sheaths of joints fringed with slender bristles, leaves much less peppery, and it grows from a rhizome; Dotted Smartweed, *P. punctatum*, is rhizomatous, has greenish flowers, and has sheaths that are not fringed but do not wrap completely around the stem

aka: *Persicaria hydropiper*

Spotted Ladysthumb
Polygonum persicaria L.

Status: FACW
Polygonaceae

Plants: 15-45 cm tall; erect or sprawling; stems reddish, hairless; from taproot

Flowers: in thick dense terminal spikes; individual flowers tiny (1.5-3.0 mm long), 5 parted

Fruit: brown to black; glossy, egg shaped; 2 mm long

Leaves: alternate; lance-shaped, 2.5-15 cm long by 0.5-1.3 cm wide, tip tapering to point; purple spot in middle of leaf resembles the mark of a lady's thumb; sheaths reddish without fringe

Habitat: marshes, streambanks, edge of ponds, ditches

Similar Species: Pennsylvania Smartweed, *P. pensylvanicum*, is similar but the sheaths at the bases of the leaves are fringed

aka: *Persicaria persicaria*

herbaceous

Arrow-Leaved Tearthumb
Polygonum sagittatum L.

Status: OBL
Polygonaceae

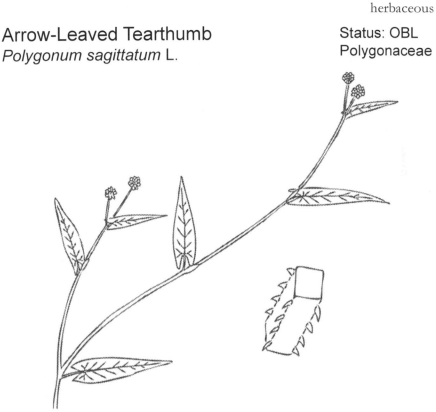

Plant: 30-90 cm long; weak stemmed and sprawling; 4 sided stems armed with backward pointing prickles; stem jointed, nodes not fringed
Flower: 5 parted, pink or white; in small tight clusters at ends of stems; 5 petals, 8 stamens
Leaves: alternate, entire; 25-75 mm lony, narrow, arrow-shaped, weak prickles below midrib
Habitat: open wet thickets, marshes
aka: *Persicaria sagittata*

Pickerelweed
Pontederia cordata L.

Status: OBL
Pontederiaceae

Plants: 30-90 cm tall
Flowers: densely arranged around terminal spike; soft blue to blue; 2 lipped, upper lip with 2 lobes, lower lip with 3 lobes with two yellow spots
Leaves: arrowhead or heart-shaped with distinctive swirling veins; glossy, deep green; up to 15 cm long, 5-10 cm across
Habitat: marshes, swamps, bogs, ponds and slow-moving rivers

Narrow-Leaved Mountain-Mint
Pycnanthemum tenuifolium Schrad.

Plant: 40-110 cm tall; leaves and stem have a strong minty smell
Flower: broad cluster of rounded flower heads; small (5-8 mm long); 5
parted: corolla whitish to pale lavender with purple spots, 2 lipped - upper
lip not lobed, lower lip 3 lobed; calyx with narrowly triangular teeth
Leaves: opposite, sessile, very narrow, pointed, up to 5 cm long by 0.4 cm
wide, not toothed; dark green above and lighter green below
Habitat: swamps, wet meadows, riverbanks
Similar Species: Virginia Mountain-Mint, *P. virginianum*, has wider leaves,
broadest near the base; usually in drier habitats

Pink Pyrola
Pyrola asarifolia Michx.

Status: FACW
Pyrolaceae

Plant: 10-40 cm tall; ; evergreen; from rhizomes

Flower: flowers with a slender spike; nodding, bell-shaped; style protrudes beyond petals; pale pink to purplish red

Fruit: 5-10 mm long, spherical capsules; 5 chambered

Leaves: leaves form basal rosette; rounded to elliptical with a heart-shaped base, 3-6 cm long, finely toothed, leather-textured, dark green and glossy on the upper surfaces and purplish beneath

Habitat: streambanks, marshes

Similar Species: the other Pyrolas have white or greenish flowers

Small Flowered Buttercup
Ranunculus abortivus L.

Plant: 15- 60 cm tall
Flower: in upper leaf axils on long stalks; tiny (3-6 mm wide); 5 parted, petals drooping, pale-yellow to yellow
Fruit: oval cluster of achenes
Leaves: heart-shaped basal leaves; 3-5 lobed stem leaves; hairless
Habitat: swamps, moist woods, roadsides
Similar Species: only Buttercup, *Ranunculus spp*, with such tiny flowers

Tall Buttercup
Ranunculus acris L.

Status: FAC+
Ranunculaceae

Plant: 30-90 cm tall; stem erect, branching, usually hairy
Flower: 5 parted, yellow; waxy, petals overlap
Fruit: achenes not hooked on end
Leaves: alternate, deeply lobed; larger leaves (4-10 cm) divided into 3-7 by deep clefts
Habitat: fields and meadows
Similar Species: Bristly Buttercup, *R. pensylvanicus*, has recurved petals, stems that are more hairy, achenes with hooked ends, and a leaf that is compound - the end segment of the leaf is separated from the others

Greater Creeping Spearwort

Ranunculus flammula L.

Status: FACW
Ranunculaceae

Plants: up to 10 cm tall; stems creeping to trailing up to 50 cm
Flowers: 5 parted, yellow; petals 4-6 mm long, as long as sepals
Fruits: 2-3 mm long, hairless, in a round cluster of 5-25
Leaves: simple, narrow and linear, 10-15 cm long, hairless
Habitat: swamps, muddy river banks
Similar Species: Water Plantain Spearwort, R. *ambigens*, is similar but has petals that are twice as long as the sepals and the leaves are slightly toothed
Notes: mildly poisonous; leaves produce acrid juice that cause skin redness, burning sensation and blisters

Bristly Buttercup
Ranunculus pensylvanicus L. f

Status: OBL
Ranunculaceae

Plant: 20- 50 cm tall; very hairy
Flower: 5 parted, yellow; petals oblong with a wider tip, recurved
Fruit: 0.8-1.2 cm long, with hooked end, hairless, in a cylindrical cluster
Leaves: deeply 3 lobed, toothed or cut
Habitat: marshes, ditches, wet meadows
Similar Species: see Tall Buttercup, *Ranunculus acris*

Yellowcress
Rorippa palustris (L.) Besser

herbaceous
Status: OBL
Brassicaceae

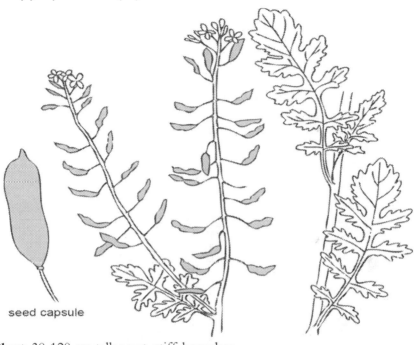

seed capsule

Plant: 30-120 cm tall; erect, stiff branches
Flower: 4 parted, yellow; small (2-3 mm) across; petals longer than sepals
Fruit: pods short and plump, spread horizontally
Leaves: alternate, lobed and coarsely toothed
Habitat: damp soils, wet shores
aka: *Rorippa islandica*

Swamp Dock
Rumex verticillatus L.

Status: OBL
Polygonaceae

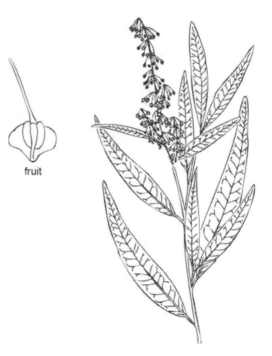

fruit

Plant: 30-100 cm tall

Flower: flowers stalked and growing in a series of whorls to form branched racemes

Fruit: seed with 3 flat wings, wings with lateral bulges and narrowing toward tip; wings entire

Leaves: alternate, entire; lower leaves lance shaped, up to 30 cm long

Habitat: swamps, wet meadows

Similar Species: Great Water Dock, *R. orbiculatus*, has larger leaves (up to 60 cm long) and more heart-shaped fruits

116

Broadleaf Arrowhead

Sagittaria latifolia Willd.

Plants: in shallow water

Flowers: on very long stalks, usually in whorls of 3; flowers 3 parted, white, 25-35 mm across

Leaves: basal, entire; most leaves arrowhead-shaped, width varies enormously; lobe of arrowhead is equal to about one half the total length of the leaf blade

Habitat: shallow water, pond edges

Similar Species: Northern Arrowhead, *S. cuneata*, is similar, but the lobes of the leaves are shorter, making up about one third of the entire length; Grassleaf Arrowhead, *S. graminea*, as very narrow lance-shaped leaves - 4-30 cm by 0.5-1.0 cm (sometimes leaves actually bladeless), and the lower flowers are on long stalks; Sessilefruit Arrowhead, *S. rigida*, has wider lance-shaped leaves, sometimes with small lobes, and the lower flowers have no stalk

Canadian Burnet
Sanguisorba canadensis L.

herbaceous
Status: FACW+
Rosaceae

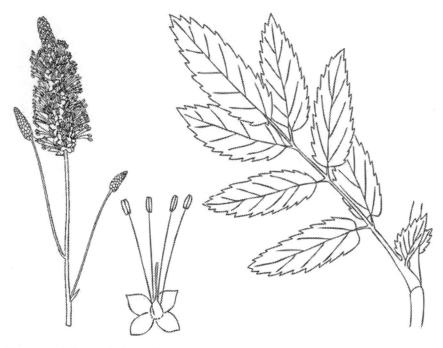

Plant: to 1.5 m tall; from rhizome
Flower: in dense spikes; white; 25-35mm across, 4 lobed petal-like calyx, no petals, long white stamens
Leaves: basal, compound with 7-17 toothed leaflets
Habitat: swamps, bogs, shores

Purple Pitcherplant
Sarracenia purpurea L.

Status: OBL
Sarraceniaceae

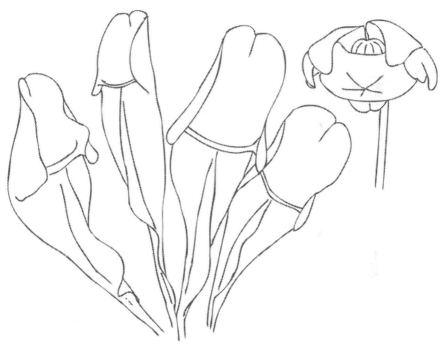

Plant: 20-60 cm tall; rosette of leaves around the base of the flower stalk; carnivorous

Flower: single, nodding, at the top of a thick leafless stalk; 5 parted, red sepals and petals with a large flattened pistil, ~5 cm wide

Leaves: pitcher-shaped, 10-30 cm long, downward pointing hairs on inside; green with brownish-red streaks and speckles

Habitat: bogs; fens

Marsh Skullcap
Scutellaria galericulata L.

Plant: 15-75 cm tall, stems squarish
Flower: bilateral, blue; 16-25 mm long; arching hooded upper lip, flaring lower lip; single flower in axil of penultimate leaves, often paired
Leaves: opposite, toothed; lance-shaped or oblong, short-stalked or stalkless
Habitat: shores, wet meadows, swampy thickets
Similar Species: Blue Skullcap, S. laterflora, has small flowers (~10 mm long) on axillary racemes
aka: *S. epilobiifolia*

Narrowleaf Blue-Eyed Grass
Sisyrinchium angustifolium Mill.

herbaceous

Status: FACW-
Iridaceae

Plant: 15-60 cm tall; flowering stem sometimes branched, distinctly winged (2-3mm), bearing a leaf-like bract in the middle

Flower: 6 parted, blue with yellow center; flower 1.9 cm across, petals with small pointed tip

Leaves: linear; up to 50 cm; grow in the shape of a fan; bright green

Habitat: swamps, wet meadows, marshes, roadsides; shores

Similar Species: Common Blue-eyed Grass, *S. montanum*, has leaves that are lighter and slightly wider leaves (over 6 mm) that reach about as high as the flower head; Needletop Blue-Eyed Grass, *S. mucronatum*, has wiry stems and very narrow, dark green, leaves (~1-2 mm) that are much shorter than the flower head. In neither of these species is the stem ever branched nor do they have a leaf-like bract

Water Parsnip
Sium suave Walt.

Status: OBL
Apiaceae

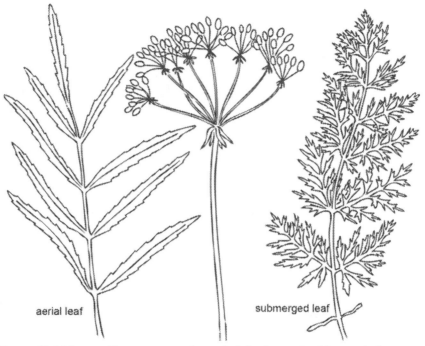

aerial leaf   submerged leaf

Plant: 60-150 cm tall; stem strongly angled, leafy toothed bracts below umbel

Flower: 5 parted, white; flowers tiny, in umbel at top of stem

Leaves: alternate, once compounded; aerial leaves with 3-7 pairs lance-shaped leaflets, sharply toothed along entire length; basal leaves, often submerged, with deeply divided leaflets

Habitat: swamps, muddy water

Similar Species: Water Hemlocks, *C. spp,* are similar but the leaves are twice or even three times compound and have coarser teeth

Bitter Nightshade
Solanum dulcamara L.

Plant: 60-200 cm long; sprawling, vine-like stems without tendrils
Flower: 5 parted, violet (sometimes white); petals recurved, protruding yellow beak formed by anthers; ~12 mm across
Fruit: cluster of egg-shaped berries, green turning red (somewhat poisonous)
Leaves: alternate, lobed (2 small lobes at base)
Habitat: moist thickets
Notes: although considered FAC- this species turns up regularly in our wetlands

Smooth Goldenrod
Solidago gigantea Ait.

Status: FACW
Asteraceae

Plants: 60-200 cm tall; stem smooth top to bottom
Flowers: yellow, tiny; each small head (~3-4 mm long) is a composite of 8-15 individual flowers; flowers in plume-like clusters at top of plant; flowers on one side of plume branch
Leaves: alternate, toothed; leaves lance-shaped, 3-veined with 2 prominent veins parallel to midrib; leaves smooth or slightly hairy, teeth fewer and duller
Habitat: moist or dry open places
Similar Species: Canada Goldenrod, *S. canadensis,* has down either on the upper stem or on the entire stem (depending on the subspecies), plus the leaves are rough and hairy; Rough-Stem Goldenrod, *S. rugosa,* has leaves with branching veins and rough hairs on the stem; many other species of Goldenrod are found in the Adirondacks

Roundleaf Goldenrod
Solidago patula Muhl. ex Willd.

Status: OBL
Asteraceae

Plants: 45-210 cm tall; angular and ridged stem
Flowers: flowers from lateral branches at top of stem; yellow; tiny; each small composite head (~3-4 mm long) is a composite of 5-12 individual ray flowers
Leaves: alternate; smaller leaves at the top and larger at the bottom; rough to the touch, ovate, toothed
Habitat: Bogs, swamps, wet meadows, shores
Similar Species: see Smooth Goldenrod, *S. gigantea*, and Bog Goldenrod, *S. uliginosa*; many other species of Goldenrod are found in the Adirondacks

Bog Goldenrod
Solidago uliginosa Nutt.

Plant: 30-150 cm tall; round and smooth stems
Flower: similar to Roundleaf Goldenrod but each composite has only 1-8 ray flowers
Fruit: smooth
Leaves: alternate; long and narrow; lance-shaped; smaller leaves at the top and larger at the bottom; rsmooth to the touch
Habitat: Bogs, swamps, wet meadows
Similar Species: see Smooth Goldenrod, *S. gigantea*, and Roundleaf Goldenrod, *S. patula*; many other species of Goldenrod are found in the Adirondacks

Branched Bur-Reed
Sparganium androcladum (Engelm.) Morong

Status: OBL
Sparganiaceae

Plant: to 1.2 m tall; leaves emergent and erect
Flower: in dense heads on branched stalks, the center branch has both male and female flower heads, the lateral branches only female flower heads; flowers tiny, greenish; stigmas 2, sometimes 1
greenish, tiny; one stigma per pistil; in dense heads, the upper heads have
Fruit: female heads develop into bur-like fruits composed of many nutlets, 15-25 mm in diameter
Leaves: alternate, entire; leaves strongly triangular in cross-section, at least up to midsection of leaf; to 1.2 m high by 5-15 mm wide
Habitat: muddy shores, shallow water
Similar Species: American Bur-Reed, *S. americanum*, also has leaves with a strongly triangular cross-section with a branched flower head, but the fruiting heads are 25-35 mm in diameter and the lateral branches of the flowering stalks have male flower heads; see also the emergent leaf form of European Bur-Reed, *S. emersum*, in the Aquatics Section

127

Broadfruited Bur-Reed
Sparganium eurycarpum Engel. ex Gray

Status: OBL
Sparganiaceae

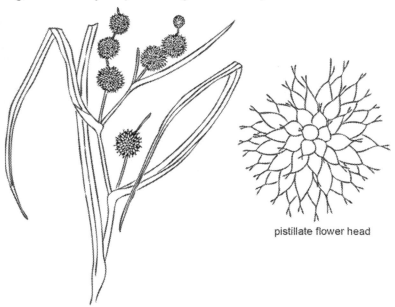

pistillate flower head

Plants: to 1.2 m tall; leaves emergent and erect
Flower:Fruit: female heads develop into bur-like fruits, 15-25 mm in diameter, composed of many nutlets
Leaves: alternate, entire; leaves nearly flat, 1.0-2.5 m long by 6-20 mm wide
Habitat: muddy shores, shallow water
Similar Species: this is the only species with 2 stigmas on the pistil; without flowers/fruits the leaves might be mistaken for Cattails, *Typha spp*

Hooded Lady's Tresses
Spiranthes romanzoffiana Cham.

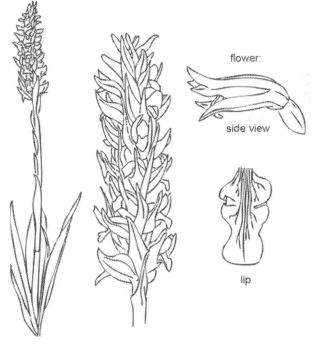

flower:

side view

lip

Plant: 10-35 m tall

Flower: spikes are tightly spiraled, usually 3 flowers per cycle of spiral; elongate ovate-lanceolate bracts; lower lip somewhat narrowed toward middle; white to creamy white, occasionally greenish

Leaves: 7-20 cm long, 0.6-1.2 cm wide, basal, linear to linear-lanceolate,

Habitat:bogs, wet meadows, streambanks

Similar Species: the very similar Nodding Lady's Tresses, *S. cernua*, has flowers less crowded on the stem, the lip distinctly narrowed toward the middle, and flowers on the spike tilted downward (nodding) - hybrids are known and are difficult to distinguish in the field; also found in the Adirondacks are Northern Slender Lady's Tresses, *S. lacera*, Shining Lady's Tresses, *S. lucida*, and Yellow Nodding Lady's Tresses, *S. ochroleuca*

Smooth Hedgenettle
Stachys tenuifolia Willd.

Status: FACW+
Lamiaceae

stem

Plant: 30-100 cm tall; 4-angled central stem with scattered hairs along the ridges; from rhizome

Flower: in small clusters in leaf axils; pink to light purple; ~10-15 mm long; upper petal lobe forming a hood; larger lower petal lobe with purple dots; may have some hairs on flowers

Leaves: opposite, narrow lance-shape to ovate-elongate; toothed outer margin; up to 13 cm long and 4 cm wide

Habitat: marshes, wet meadows, streambanks, ditches

Notes: two variant forms used to be considered separate species: Smooth Hedgenettle and Rough Hedgenettle

New York Aster
Symphyotrichum novi-belgii (L.) Nesom

Status: FACW+
Asteraceae

Plant: 30-120 cm tall; stem smooth to slightly downy with waxy bloom
Flower: heads a composite of violet ray flowers (25-50) and yellow disk flowers, 2-4 cm wide; head surrounded by reflexed floral bracts
Leaves: alternate; narrow lance-shaped (7 or more times as long as wide, to 7 cm long) leaves; no leafstalks but do not clasp around the stem; usually toothless
Habitat: meadows, shores, wet thickets, swamps
Similar Species: New England Aster, *S. novae-angliae*, has hairy stems with somewhat wider leaves (~6 x long as wide) that clasp part way around the stem, toothless; flower heads somewhat wider (2-5 cm), ray flowers deeper purple, more numerous (45-100); floral bracts sticky
aka: *Aster novi-belgii*

Smooth White Old Field Aster

Symphyotrichum racemosum (Ell.) Nesom

Status: FACW
Asteraceae

Plants: 60-150 cm tall; smooth purple stems

Flowers: many small (less than 12 mm across) crowded on upper stems of plant; flower heads with white ray flowers surrounding yellow to reddish disk flowers

Leaves: alternate; toothed (at least the lower ones); slender; no leafstalk; often with small leaflets in the axils of larger leaves

Habitat: fields and wet meadows

Similar Species: White Panicle Aster, *Symphyotrichum lanceolatum* aka *Aster simplex*, has slightly larger flowers (more than 12 mm across) and leaves that usually lack teeth and also lack the small leaflets in the axils

aka: *Aster vimineus*

Skunk Cabbage
Symplocarpus foetidus Willd.

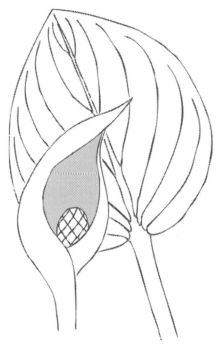

Plant: 30-60 cm tall; flowers emerge early spring, before leaves
Flower: hooded spathe knob-shaped, 7-14 cm tall, mottled purple brown to green; spadix ball-like, surrounded by spathe
Leaves: basal, coiled at first then unfolding; up to 60 cm long and 30 cm wide; ovate to cordate, rounded at the base; mature leaves have a foetid odor when crushed
Habitat: swamps, marshes

Tall Meadow Rue
Thalictrum pubescens Pursh

Status: FACW+
Ranunculaceae

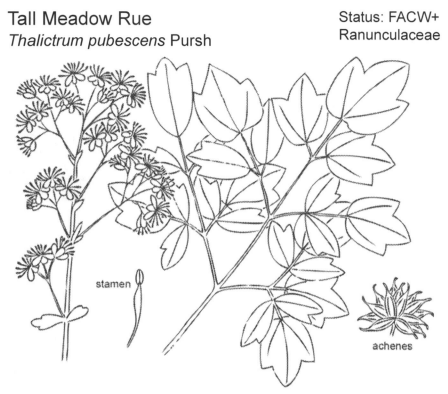

stamen

achenes

Plant: 1.0-2.5 m tall

Flower: flowers in plumes mostly unisexual; at top of stalk; white; 4 parted, petals lacking, 4 whitish sepals (usually shed early); long white club-shaped erect stamens

Fruit: achenes ellipsoidal, 4-6 mm; in ball-shaped head

Leaves: alternate, compound; divided and subdivided into leaflets, each with 2-3 lobes; light green

Habitat: sunny swamps and streamsides, wet meadows

Similar Species: Rue Anemone, *T. thalictroides* aka *Anemonella thalictroides*, has a single white to pinkish flower, 1.5-2.0 cm across, with 5-10 petal-like sepals on a small head; leaves are whorled, compound (three leaflets), located below flower head and on long separate stalks

Foamflower
Tiarella cordifolia L.

herbaceous

Status: FAC-
Saxifragaceae

Plant: 15-30 cm tall
Flower: 5 parted, white; long conspicuous stamens
Leaves: basal, toothed and lobed; deeply heart-shaped base with 5-7 shallow lobes
Habitat: rich woods
Similar Species: Naked Miterwort, *Mitella nuda*, has leaves that although toothed are more rounded and petals that are fringed; Early Saxifrage, *Saxifraga virginiensis*, has the basal leaves without leafstalks and flowers with bright yellow stamens in branched clusters

Marsh St John's Wort
Triadenum virginicum (L.) Raf.

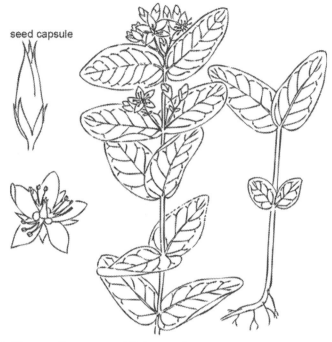

seed capsule

Plant: 30-45 cm tall; stems reddish purple
Flower: 5 parted, pink; 12-18 mm across; 3 groups of 3 stamens alternate with 3 large orange glands; flowers in small clusters, either terminal or in upper leaf axils; sepals purplish
Leaves: opposite; oblong to egg-shaped, very blunt, not toothed; midvein reddish purple; leafstalks lacking
Habitat: bogs, swamps
aka: *Hypericum virginicum*

White Trillium
Trillium grandiflorum (Michx.) Salisb.

Plant: 15-50 cm tall
Flower: 3 parted, white often turning pale pink with age due to a fungal
infection; petals 4-6 cm long; erect flowerstalk
Leaves: single whorl of 3; leaves symmetrical, wide; branched veins
Habitat: rich woods, floodplains
Similar Species: White Trillium, *T. grandiflorum*, is the only species with
an entirely white flower on an erect flower stalk; if flowers are not present
Painted Trillium, *T. undulatum*, can be distinguished by its stalked leaves;
Nodding Trillium, *T. crenuum*, usually shows the remains of the drooping
flower stalk; Red Trillium, *T. erectum*, usually grows as scattered individuals
and has somewhat narrower leaves whereas White Trillium usually grows in
large beds

Wide Leaved Cattail
Typha latifolia L.

T. angustifolia

staminate flowers

pistillate flowers

T. latifolia

Plant: 1.2-2.7 m tall

Flower: tiny, brown; flowering stalk stiff, topped with dense cylindrical spikes; females flowers below, spike about 25 mm in diameter, male flowers above, the two touching or up to 4 cm apart

Fruit: mature "cattail"

Leaves: entire, alternate (may appear to be basal); widest 10-23 mm

Habitat: marshes, wet ditches

Similar Species: *T. angustifolia* has narrower leaves (5-11 mm), a thinner female spike, and a wider gap between the male and female flowers; a hybrid between *T. angustifolia* and *T. latifolia* called *Typha × glauca* is present but can be reliably distinguished only by using DNA technology

Stinging Nettle
Urtica diocia L.

Plant: 60-100 cm tall; stem and leaves densely covered with stinging hairs; stem hollow, 4 angled
Flower: green-brown, tiny; in clusters on two spikes arising in each of the upper leaf axils; male and female flowers may be on different plants
Leaves: opposite, coarsely toothed; oval; covered with stinging hairs
Habitat: damp soil, thickets
Similar Species: Wood Nettle, *Laportea canadensis*, also has stems covered with stinging hairs but the leaves are alternate

herbaceous

Cranberry
Vaccinium macrocarpon Aiton

Status: OBL
Ericaceae

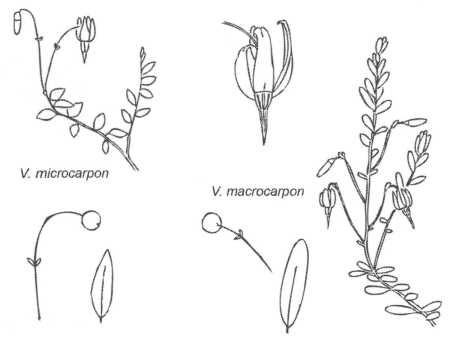

V. microcarpon

V. macrocarpon

Plants: 10-25 cm tall; creeping evergreen; woody stems
Flowers: single, from leaf axils; ~12 mm long; pale pink to white; petals recurved, stamens form beak; pair of small bracts about two-thirds of the way along the flower stalk
Fruits: large round berries, red, 10-15 mm across
Leaves: alternate, simple; 15 mm long; oval, rounded tips; leaf margins slightly rolled under; dark green, paler below
Habitat: bogs
Similar Species: Small Cranberry, *V. oxycoccos*, has thinner stems and smaller leaves with more acute tips, white underneath; flowers come from ends of stems; bracts on flower stalk about half way along; fruit 6-12 mm across

False Hellebore
Veratrum viride Aiton

Status: FACW+
Liliaceae

Plants: 60-200 cm tall; root very poisonous
Flowers: in large (20-50 cm tall) branched cluster at top of stem; flower 6 parted, yellowish-green then green, anthers dark
Leaves: alternate, entire; large, oval, clasping, conspicuous parallel ribs; 20-30 cm long and 7-15 cm broad
Habitat: swamps, wet woods

141

Common Vervain
Verbena hastata L.

Status: FACW+
Verbenaceae

leaf variations

Plant: 50-150 cm tall; stem somewhat squarish
Flower: 5 parted; blue-violet; thin spikes of small (5-9 mm across) flowers bloom sequentially, advancing toward the tip
Fruit: 0.25 cm long, nutlets 4
Leaves: opposite, toothed; lower leaves lance-shaped, sometimes lobed; coarsely toothed, short-stalked
Habitat: wet meadows, riparian, roadsides, railroads, streambanks
Similar Species: Narrow-Leaved Vervain, *M. simplex*, has narrow lance-shaped leaves with fewer teeth that gradually taper toward the base (no distinct leaf stalk); Marsh Vervain, *V. urticifolia*, has broader leaves and less dense flowers on the spikes and with white corollas

Marsh or Narrow-Leaved Speedwell
Veronica scutellata L.

Status: OBL
Scrophulariaceae

Plant: 11-26 cm tall
Flower: bilateral, blue-violet; 4 petals, upper wider, lower narrower, than two lateral; 2 stamens; ~ 7 cm across; in loose racemes in axils of leaves
Fruit: capsules, heart-shaped, flat, notched
Leaves: opposite, smooth edged (or with a few tiny teeth), slender, 25-75 mm long
Habitat: swamps, marshes
Similar Species: Water Speedwell, *V. anagallis-aquatica* aka *V. catenata*, has wider leaves that clasp the stem; its flower racemes are longer and contain more flowers

Sand Violet
Viola affinis Leconte

Status: FACW
Violaceae

Plant: 15-25 cm tall; long stems from base; thick rhizome
Flower: violet to light violet with white throat; petals spread open; hairs on lower petal
Leaves: heart-shaped leaves; glossy; coarsely toothed
Habitat: riparian, wet meadows
Similar Species: Sand Violet, *V. affinis*, has a distinctive combination of characters: no leaves on stems, thick rhizome, petals spread open, and long hairs on lower petal; many other species of blue flowered violets are found in the Adirondacks: Hooked-Spur Violet, *V. adunca*; Dog Violet, *V. conspersa*; Marsh Blue Violet, *V. cucullata*; Alpine Violet, *V. labradorica*; Northern Bog Violet, *V. nephrophylla*; New England Blue Violet, *V. novae-angliae*; Long-Spurred Violet, *V. rostrata*; Arrow-Leaved Violet, *V. sagittata*; Great-Spurred Violet, *V. selkirkii*; Northern Blue Violet, *V. septentrionalis*; Wooly Blue Violet, *V. sororia* (aka Common Blue Violet, *V. papilionaceae*). All of these seem to get into wetland habitats at times.

Sweet White Violet
Viola blanda Willd.

Plant: 7-25 cm tall; perennial; spreading with runners
Flower: 5 parted, white; lower petals with purple veins and point forward; upper petals twisted and bent backward.
Fruits: oval shape; purplish color.
Leaves: Oval with heart-shaped base
Habitat: shady moist swamp forests
Similar Species: Northern White Violet, *V. macloskeyi* aka *V. pallens*, has flowers where the upper petals are not twisted forward; Kidney-Leaved Violet, *V. renifolia*, has a kidney-shaped leaf and downy stems; Canada Violet, *V. canadensis*, has leaves on the leaf stalk; the yellow flowered violets found in the Adirondacks are Downy Yellow Violet, *V. pubescens*, and Round-Leaved Yellow Violet, *V. rotundifolia*

Barren Strawberry
Waldsteinia fragarioides (Michx.) Tratt.

Status: Unknown
Rosaceae

Plant: 8-20 cm tall; not from runner
Flower: 5 parted, yellow; in small clusters on separate stalk from leaves
Fruit: dry and inedible, not a berry
Leaves: basal, compound; base of leaflets wedge-shaped, irregularly toothed on end; leaf stalks slightly hairy
Habitat: woods, clearings
Similar Species: Strawberries, *Fragaria spp.* arise from runners, have white flowers, berry-like fruits, and more elliptical leaflets; Three Leaf Goldthread, *Coptis trifolia*, has smooth leaf stalks; its flowers (if present) are single and white

Northern Yelloweyed Grass
Xyris montana Ries.

Plant: 5-30 cm tall; perennial
Flower: from cone-like head at top of stalk; 3 parted, yellow, 6-7 mm across; open in the morning
Leaves: mostly basal, long and linear (4-15 cm by 1.0-2.5 mm); deep green
Habitat: sphagnum bogs, tamarack swamps
Similar Species: Bog Yelloweyed Grass, *X. difformis*, has longer and wider leaves (10-50 cm by 5-15 mm) and is found in wet sandy places, often swampy river edges

Golden Alexanders
Zizia aurea (L.) W. D. J. Koch

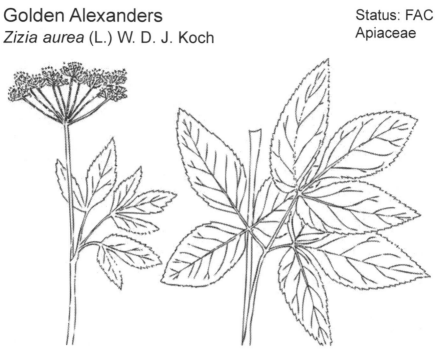

Plant: 30-70 cm tall; stem branching, often red-tinged
Flower: yellow, 5 parted; tiny, arranged in umbels
Leaves: alternate compound; divided into 3's and subdivided again into 3-7 leaflets; individual leaflets lance-shaped to egg-shaped, with sharp teeth
Habitat: meadows, wet thickets, swamps
Similar Species: Wild Parsnip, *Pastinaca sativa*, also with yellow flowers is larger and has leaves divided into 5-15 sessile toothed leaflets

Aquatic Plants

Aquatic plants are plants that can only grow in water or in soil that is inundated/saturated with water. They have developed special adaptation to live in aquatic environments. They can either float on water surface or can be submerged under water. In this book, we divide aquatic plants into three groups. The three groups are separated below, but in some cases there may be overlap or it may be difficult to tell if a specimen is "attached" to a muddy surface or merely "resting" on it.

1. **Floating Leaves Only**: floating leaves either on or under the water *page 150*
2. **Submerged & Floating Leaves**: plants with leaves that both float on the surface of the water or are submerged; the plants are rooted or somehow attached to the bottom *page 155*
3. **Submerged Leaves Only**: plants with leaves that are submerged; the plants are rooted or somehow attached to the bottom *page 159*

Water Shield
Brasenia schreberi J.F. Gmel.

floating leaves only
Status: OBL
Cabombaceae

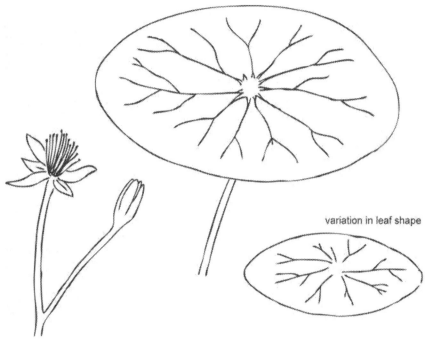

variation in leaf shape

Plant: floating leaves
Flower: dull purple, 3 (sometimes 4) parted; ~20 mm across
Leaves: floating; alternate; oval shield-shaped, 5-10 cm long, green in summer then becoming purplish-red; long stalk attached to underside at center of leaf; stalk and underside of leaves covered with gelatinous slime
Habitat: ponds, sluggish streams
Similar Species: shield-shaped leaves are distinctive

Lesser Duckweed
Lemna minor L.

Status: OBL
Lemnaceae

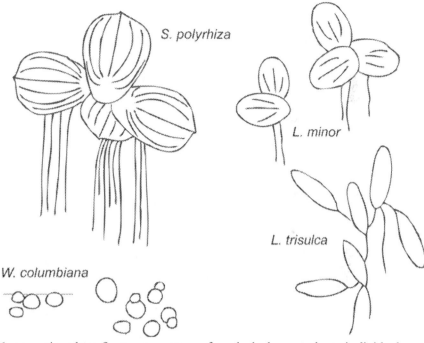

S. polyrhiza

L. minor

L. trisulca

W. columbiana

Plant: entire plant floats on water surface, lacks leaves; plants individual or in small rosettes; each individual plant with a single root from lower surface; plants elliptical, symmetrical, with 3 obscure veins; 1.0-2.5 mm long; green or slightly reddish-purple

Flower: rarely flower

Habitat: quiet waters, mesotrophic to eutrophic

Similar Species: Star Duckweed, *Lemna trisulca*, is significantly longer than wide; Duckmeat, *Spirodella polyrhiza*, has multiple roots on each frond: Columbian Watermeal. *Wolffia columbiana*, is small (~1.0 mm long), without roots, and floats at or just beneath the water surface

151

Yellow Pond Lily
Nuphar lutea (L.) Sm.

floating leaves only

Status: OBL
Nymphaceae

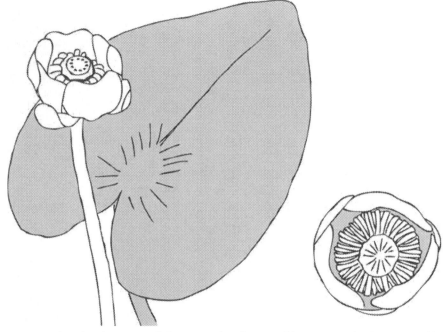

Plants: floating leaves and flowers arise from a rhizome on the bottom off the pond

Flower: 6 parted, yellow; 25-50 mm diameter; sepals green to yellow, often with maroon near base; stamens and stigmatic disk yellow or reddish

Leaves: alternate, entire; usually floating, sometimes submerged; notched with rounded waters lobes; blade 7-35 cm, length 1.2-1.6 x width; sinus 1/3 to 1/2 length; leafstalk flattened with medial ridge, wings along margins

Habitat: ponds, lakes, sluggish streams, ditches

Similar Species: this species used to be divided into several species, three of which occur in the Adirondacks: *Nuphar microphylla*, *N. rubrodisca* and *N. variegata*. These are now considered subspecies of *N. lutea*

aka: Yellow Water-Lily and Spatterdock

White Water Lily
Nymphaea ordorata Aiton

floating leaves only

Status: OBL
Nymphaceae

Plant: floating leaved, from rhizome

Flower: white ; 6-19 cm across, open during the day; sepals and numerous (7-43) petals in whorls of 4; sepals green to reddish green; petals white, lanceolate and tapering toward tip; stamens 35-120, yellow; stigmatic disk yellow

Leaves: basal, entire; ovate to nearly circular, notch with sharp often eared corners; 10-40 cm long; green above, greenish to reddish purple below

Habitat: ponds, lakes, sluggish streams, marshes, ditches, sloughs, canals

SubSpecies: divided into two subspecies, both of which are found in the Adirondacks: *Nymphaea odorata odorata* has leaves with undersides usually reddish to reddish purple and unstriped leaf stalks; *Nymphaea odorata tuberosa* has leaves with undersides greenish to light reddish purple, leaf stalks with distinct purple-brown stripes and an enlarged tuberous rhizome

Little Floating Heart
Nymphoides cordata (Ell.) Fern.

floating leaves only

Status: OBL
Menyanthaceae

Plant: leaves floating or stranded later in season
Flower: 5 parted, white; 12-18 mm across; flowers in umbel; some flowers replaced by small tuberous roots
Leaves: basal, entire; heart shaped, long stalked; 3-7 cm long
Habitat: ponds, shallow still water
Similar Species: leaves much smaller than either White Water Lily or Yellow Pond Lily

Floating Pondweed
Potamogeton natans L.

Status: OBL
Potamogetonaceae

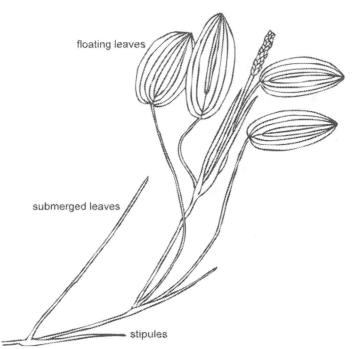

floating leaves

submerged leaves

stipules

Plant: floating and submerged leaves
Flower: 4 parted, born on spikes
Fruit: drupe
Leaves: alternate, entire; submerged leaves with stipule not fused to stem (attached only at base), linear, ribbon-like, no more than 2 mm wide; floating leaves present, 1.5-12.0 cm long; often somewhat heart-shaped at base in still water, otherwise more elliptical, leafstalk pale at junction with blade
Habitat: ponds, shallow waters, acid to alkaline waters
Similar Species: there are about 25 species of *Potamogeton* in the Adirondacks. Some have floating leaves, some have submerged leaves, some have both; you will need to collect a specimen for a botanist to identify if you need to know which species you have

155

Yellow Water Crowfoot
Ranunculus flabellaris Raf.

Plant: 15-60 cm long; floating aquatic, though may come to rest on muddy surfaces later in season; stem hollow

Flower: 5 parted, waxy yellow

Leaves: alternate, compound; leaves below water divided and subdivided into filamentous lobes; leaves above water less subdivided, not as filamentous

Habitat: quiet ponds

Similar Species: the only aquatic *Ranunculus* with yellow petals; Longbeak buttercup, *R. longirostris*, has white petals

Water Chestnut
Trapa natans L.

Status:OBL
Trapaceae

Plant: floating annual plants
Flower: four petaled white flowers
Fruit: nuts with four barbed spines
Leaves: two types of leaves; feather-like submerged leaves along the stem; floating saw-tooth edges, ovoid or triangular leaves with inflated petioles.
Habitat: ponds, lakes, streams, shallow slow moving waters

Greater Bladderwort
Utricularia macrorhiza LeConte

submerged and floating leaves

Status: OBL
Lentibulariaceae

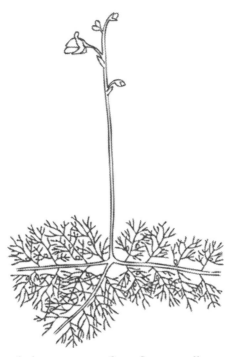

Plant: leaves float below water surface, flower stalk extends above water

Flower: yellow, bilateral; 1-2 cm long; 2 lips, upper and lower equal size, with a forward-facing spur about 2/3 the length of the lower lip; stalk 6-20 cm tall with 6-20 flowers

Leaves: finely divided but with central rachis; bear tiny bladders which trap insects

Habitat: shallow waters, ponds

Similar Species: Several other species of Bladderworts occur in the Adirondacks: Horned Bladderwort, *Utricularia cornuta*; Hiddenfruit Bladderwort, *U. geminiscarpa*; Humped Bladderwort, *U. gibba*; Flatleaf Bladderwort, *U. intermedia*; Lesser Bladderwort, *U. minor*; Eastern Purple Bladderwort, *U. purpurea*; Little Floating Bladderwort, *U. radiata*; and Lavender Bladderwort, *U. resupinata*

Coontail
Ceratophyllum demersum L.

submerged leaves only

Status: OBL
Ceratophyllaceae

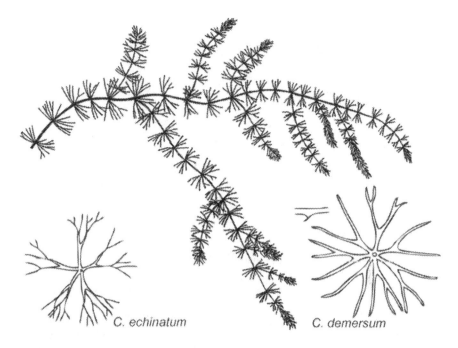

C. echinatum   C. demersum

Plant: submerged leaf aquatic with densely bushy stem tips
Leaves: whorls of dichotomously branching (usually branch no more than twice) leaves; margins of leaves have tooth-like serrations, each arising from a fleshy base
Habitat: quiet waters; ponds, lakes, swamps, marshes, streams, ditches
Similar Species: Prickly Hornwort, *Ceratophyllum echinatum,* has leaves that branch 3-4 times and teeth without fleshy base

Canadian Water Weed
Elodea canadensis Michx.

submerged leaves only

Status: OBL

Hydrocharitaceae

Plant: submerged leaves; plant often reproduces from fragments of stem; free-floating or rooted

Flower: female flowers on thread-like flowerstalks that arise from submerged spathe on stem; female flowers float on surface, 3 pinkish or white petals; male flowers produced in submerged spathes on stem, male flowers are released, then float to surface

Leaves: usually in whorls of 3, sometimes 2; ovate leaves, ~2mm wide, finely (one cell) toothed; either limp or somewhat rigid, often encrusted with epiphyic algae and/or lime

Habitat: lakes, ponds, shallow waters

Similar Species: Nuttall's Water Weed, *E. nuttallii*, has narrower (~1.3 mm) and limper leaves

160

Whorl-Leaf Watermilfoil
Myriophyllum verticillatum L.

submerged leaves only

Status: OBL
Haloragaceae

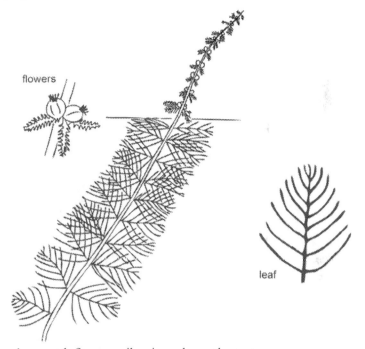

flowers

leaf

Plant: submerged, flower spike rises above the water
Flower: four petals, occur in the leaf bases on the emergent stalks, bisexual, male flowers at the top of the spike, female flowers below
Fruit: small fruit splits into 4 chambers with each containing one seed
Leaves: two types of leaves; the submersed feather-like leaves arranged in whorls around the stem with 4-5 leaves per whorl, each leaf with 7-17 paired leaflets; the emergent leaves are arranged in whorls around the stalk
Habitat: lakes, ponds, ditches, small streams, shallow waters
Similar Species: easily confused with the many other Watermilfoil in the Adirondacks: Alternate-Flowered Watermilfoil, *Myriophyllum alterniflorum*; Farwell's Watermilfoil, *M. farwellii*; Low Watermilfoil, *M. humile*; Cut-Leaf Watermilfoil, *M. pinnatum*; Common Watermilfoil, *M. sibiricum*; European Watermilfoil, *M. spicatum*; and Slender Watermilfoil, *M. tenellum*

161

Nodding Waternymph
Najas flexilis (Willd.) Rostk. & Schmidt

Status: OBL
Najadaceae

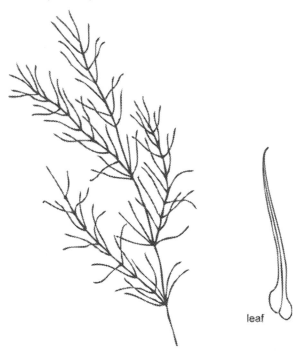

leaf

Plant: submerged annual plants, opposite leaves; leaves often clustered near the tips of the stems
Flower: tiny (2-3 mm), in clusters at the base of the leaves; male and female flowers occur separately on the same plant
Leaves: glossy, green, finely toothed, long and narrow (1-3 cm long and 1-2 mm wide)
Habitat: ponds, lakes and sluggish streams, fresh to brackish water
Similar Species: Nodding Waternymph's green glossy leaves distinguish this species from the brownish Slender Waternymph, *N. gracillima*
aka: Slender Water Nymph or Slender Naiad

Small Pondweed
Potamogeton pusillus L.

submerged leaves only

Status: OBL
Potamogetonaceae

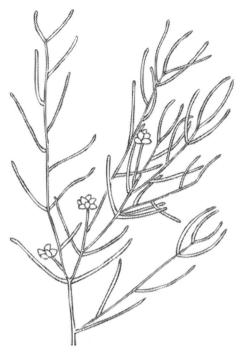

Plant: perennial with submerged leaves
Flower: in 1-4 whorls on spikes
Fruit: 1.5-3 mm long, rounded back, straight beak
Leaves: alternate, stalkless, narrow, linear, 2-7 cm long, with pointed to rounded tips and 3 veins.
Habitat: ponds, lakes, streams, shallow waters, acid to alkaline waters, fresh to brackish waters
Similar Species: Floating Pondweed, *P. natans (page 155)*

Narrowleaf Bur-Reed
Sparganium angustifolium Michx.

submerged leaves only

Status: OBL
Sparganiaceae

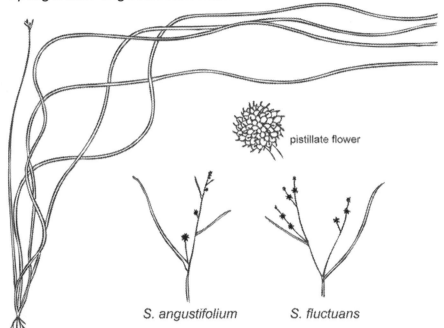

pistillate flower

S. angustifolium   S. fluctuans

Plants: submerged leaves, floating, or emergent to 1 m long, usually not branched

Flowers: usually does not flower; if present, emergent on stems; female flower heads are spiny, round balls that turn from green to brown as they mature; several small male flower heads are located above female heads

Fruits: 1-3 cm diameter, hard, dry cluster in reddish to brownish

Leaves: slender (less than 5mm) and long (up to more than 2 m long); rounded on back; upper leaves and flower bracts dilated at base

Habitat: ponds, shallow waters, acid to alkaline waters

Similar Species: Floating Bur-Reed , *S. fluctuans*, is similar but has leaves greater than 5 mm wide that are not rounded on the back; has upper leaves and flower bracts that are not dilated at the base; the European Bur-Reed, *S. emersum* (aka *S. chlorocarpum*), has both floating and emergent leaves; erect leaves are keeled, 4-10 mm wide, keeled but not strongly triangular; floating leaves are similar but 4-18 mm wide

American Eelgrass
Vallisneria americana Michx

submerged leaves only

Status: OBL

Hydrocharitaceae

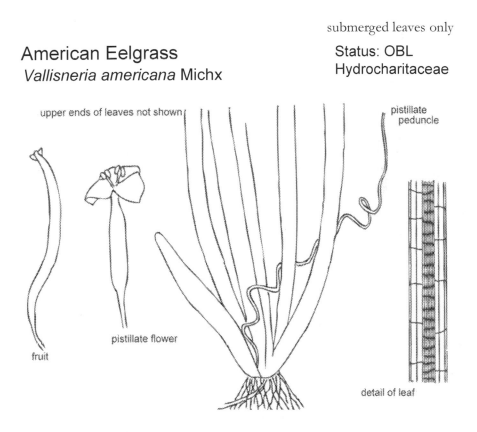

upper ends of leaves not shown

pistillate peduncle

pistillate flower

fruit

detail of leaf

Plant: submerged perennial
Flower: male and female flowers on separate plants; tiny male flowers of 1 mm at the plant base; male flowers are released, then float to surface; female flowers have 3 small white or transparent petals from a tube-like sheath, 2-3 cm long.
Fruit: cylindrical,slightly curved pods, 5-10 cm long
Leaves: flat, thin and ribbon-like, midrib consists of many parallel veins bounded by clear outer zones.
Habitat: ponds, lakes, and quiet streams, shallow waters
Similar Species: easily distinguished from the species of Bur-Reed, *Sparganium spp*, with submerged leaves, which do not have the distinctive midrib in the leaf
aka: American Wild Celery, Water Celery, and Tapegrass

Index

166

169